BIOCHEMISTRY RESEARCH TRENDS

LACTOSE

STRUCTURE, FOOD INDUSTRY APPLICATIONS AND ROLE IN DISORDERS

BIOCHEMISTRY RESEARCH TRENDS

Additional books in this series can be found on Nova's website
under the Series tab.

Additional e-books in this series can be found on Nova's website
under the e-book tab.

BIOCHEMISTRY RESEARCH TRENDS

LACTOSE

STRUCTURE, FOOD INDUSTRY APPLICATIONS AND ROLE IN DISORDERS

DAVID GREEN
AND
EMMA LEE
EDITORS

NOVA BIOMEDICAL
New York

For permission to use material from this book please contact us:
Telephone 631-231-7269; Fax 631-231-8175
Web Site: http://www.novapublishers.com

Additional color graphics may be available in the e-book version of this book.

Library of Congress Cataloging-in-Publication Data

ISBN: 978-1-62808-198-5

Library of Congress Control Number: 2013941201

Published by Nova Science Publishers, Inc. † New York

Contents

Preface

Lactose is a disaccharide which is the main carbohydrate present in the milk of most mammals, including humans. In this book, the authors present current research in the study of the structure, food industry applications and role in disorders of lactose. Topics include the microbial transformation of lactose and the potential of Β-galactosidases for probiotic and prebiotic purposes; the evolution and spread of the dominant genetic trait responsible for continued adult digestion of lactose; whey lactose fermentation by kefir immobilized on agro-industrial wastes; the tableting properties of crystalline and amorphous lactoses and their application in the production of pharmaceutical compacts; and lactose and its influence in the diffusion of divalent metal chlorides.

Chapter 1 – In the last decades, the interest of food industry has focused in the development of functional foods because of the high demand of healthy foods by consumers around the world. In this sense, foods containing probiotic microorganisms have received particular attention. One of the firsts beneficial properties ascribed to probiotics is their ability to contribute to the intestinal metabolism of lactose from the diet and the amelioration of intolerance symptoms. This effect is mediated by the enzyme β-D-galactosidase of the microorganisms included in the product that can hydrolyze the disaccharide into the food before consumption, as well as to contribute to the intraintestinal digestion of lactose.

However, β-galactosidases have more biotechnological potentialities with a wider spectrum of application. This enzyme possess, depending on the reaction conditions (substrate concentration, aw, pH, reaction temperature), hydrolase or transferase activities, owing the last one to the production of oligosaccharides containing 2 to 10 monomers obtained by transgalactosylation reactions that transfer galactose hydroxyl groups to the disaccharide lactose and/or the different oligomers that are being generated.

Galactooligosaccharides (GOS) are well known prebiotics that modulate the colonic microbiota in a healthy manner and also have other promising industrial applications such as their addition to foods as no-cariogenic, low-calories sweeteners. The raw material used for GOS production could be the lactose from milk as well as the whey discarded by dairy industries from cheeses manufactures. Since whey is considered an environmental contaminant its utilization in this regard seems to be an attractive option.

In the present chapter, the potential of well-known probiotics (such as lactic acid bacteria, bifidobacteria, propionibacteria and yeasts) to contribute to the intestinal metabolism of lactose and/or to produce prebiotic GOS by their β-galactosidase activities is reviewed. In many cases the main biochemical properties of the enzyme of the microorganisms with the higher hydrolytic activities have been determined as well as their abilities to improve lactose digestion in animal models and clinical trials. In this regard, multiple therapeutic strategies have been assessed with different results.

On the other side, the production of GOS by transgalactosylation using *β*-galactosidases from GRAS (Generally Recognized As Safe) microorganisms has also been studied. Depending on the enzyme source and the reaction conditions, the extent of oligosaccharide formation could be as high as 40% of the total sugar content in the solution and three to eleven types of oligosaccharides could be found, including disaccharides other than lactose, trisaccharides and various tetrasaccharides. Although some suitable sources of β-galactosidases that can be used to synthesize novel GOS have been characterized, further studies on the prebiotic effects of these GOS in pure cultures of bifidobacteria, and colonic microbiota of animals and human volunteers are still missing.

Besides, the selection of probiotic microorganisms with appropriate β-galactosidase activities should be complemented with in vivo assays (survival and activity in the intestinal conditions) and certain technological parameters in order to be included in the formulation of novel functional foods, particularly fermented products enriched in GOS and β-galactosidase.

Chapter 2 – Lactose is a disaccharide which is the main carbohydrate present in milk of most mammals, including humans. At birth mammals possess lactase for digestion but lose this enzyme after weaning. In humans however, about 1/4 to 1/3 of the world's population retain lactase throughout adulthood, allowing continued consumption of milk.

The consequences of this human divide are the incurring of digestive symptoms in those who have limited intestinal lactase and consume large enough quantities at a time.

The medical issues arising from this dichotomy then include the consequences of lactose associated symptoms and the more important consequences of diary food avoidance. In addition dairy foods are linked or hypothesized to increase or decrease risks for many diseases. Perhaps one of the most interesting aspects of the world lactase dichotomy is the relationship of population distribution and geographic patterns of modern day, so called "western" lifestyle linked diseases. The asymmetric distribution and increasing incidence/prevalence of these diseases in developing nations sheds light on the possible co evolutionary impact of lactase pheno or genotype on such diseases. The development of the asymmetric lactase patterns and consequent medical aspects will be discussed.

Chapter 3 – Whey is the main liquid by-product of the dairy manufactures, in which a large quantity of the milk lactose is removed. So, whey consists of lactose (4.8-5.3%), proteins (0.8-1.6%), fats (0.3-1.3%), vitamins (0.8%) and minerals. The fat and protein components are rather easily recovered, but lactose responsible for 90% of whey's COD remains unused.

The last two decades, the natural mixed culture kefir was chosen for whey lactose exploitation, because its microbiological composition ensures simultaneous alcoholic and lactic acid whey fermentations resulting to useful products formation like ethanol, lactic and propionic acid. Also, the immobilization of kefir cells on various natural food grade supports and the preparation of biocatalysts suitable for easy and fast whey fermentation have received considerable attention by researchers.

In this chapter the whey lactose fermentation kinetics by kefir immobilized on Delignified Cellulosic Materials (DCM), Gluten Pellets (GP), Brewer's Spent Grains (BSG), Malt Spent Rootlets (MSR) and Corn Starch Gel (CSG) was examined in relation to free kefir cells (FC) for comparative reasons. Also, lactose uptake rate by immobilized and free kefir cells during fermentation was monitored by adding ^{14}C-labelled lactose in whey and the produced ethanol was determined.

The biocatalysts prepared by immobilization of kefir cells on agro-industrial wastes were proved effective for whey fermentation at 30°C and pH 5.5, reducing significantly fermentation time and increasing lactose uptake rate in comparison to free cells. Specifically, using cheap, abundant and food grade BSG or DCM biocatalysts for whey fermentation is potentially acceptable by industries as the fermentation time was estimated only about 7 hours in that case. The highest alcohol concentration was observed by MSR biocatalyst.

Chapter 4 – Lactose is the most widely used excipient for the manufacture of tablets and capsules due to its relatively low cost, high solubility, good absorption and rapid metabolism in the human body. It is blended with a drug and other excipients and compressed to make tablets for oral administration. However, lactose is not desirable to formulate alkaline drugs or active ingredients with amine moieties due to the formation of a Maillard reaction. It is available as α-lactose monohydrate, anhydrous α-lactose, anhydrous β-lactose and spray-dried lactose.

Spray-dried lactose had the best flow. On the contrary, β-lactose exhibited the poorest flow, but along with the anhydrous α-lactose presented the highest binding ability. Conversely, α-lactose monohydrate presented the poorest compactibility, the fastest disintegration time and its compactibility was not affected by moisture.

Spray-dried lactose is composed of α-lactose monohydrate crystals glued together by amorphous lactose. On the contrary, α-lactose and α/β-lactose were the most fragmenting materials. In turn, these materials were the least sensitive to magnesium stearate. Lactose tablets independent of the crystalline form showed a moisture-dependent tensile strength resulting into a decreased compactibility. However, under storage compacts showed a very small elastic recovery.

It is concluded that compactibility and compressibility of lactose was related to its polymorphic form.

Chapter 5 – Lactose (disaccharide derived from the condensation of galactose and glucose) is not only a technological important compound, but it also enjoys biological, medical, pharmaceutical, food and biomedical applications. As a typical non-electrolyte carrying hydrophilic hydroxyl groups capable of hydrogen-bonding, its properties play a significant role in the reaction conditions of many current industrial processes, such as enzymatic conversion of biomass to useful chemicals, and in many biochemical interactions with metal ions.

However, while numerous studies have been carried out on the thermodynamic properties of binary aqueous lactose solutions, data are more limited on the transport behaviour of these sugar systems in aqueous solutions. The characterization of the transport properties, particularly diffusion coefficients, is important for fundamental reasons, helping us to understand the nature of the structure of the aqueous systems containing metal ions, and for practical application in fields such as corrosion. In fact, they provide a direct measure of molecular mobility, an important factor in the preservation of biological materials in sugar matrixes.

Lactose is the most important carbohydrate present in the milk of most mammals. As a consequence, for the lactose-intolerant persons it has been verified that the ingestion of enough calcium may be difficult. For these people, who reveal inability or insufficient ability to digest lactose, because of a deficiency of the enzyme lactase, it is often recommended that they have calcium supplements, such as calcium carbonate or others.

Hopefully, the studies reported here will lead to an increase in the know-how, which will allow a better understanding of the physicochemical conditions underlining the diffusion phenomena occurring in different biological systems. Thus, we proposed to extend our studies concerning the experimental determination of the diffusion coefficients for systems containing lactose, that is, for aqueous system of calcium chloride plus lactose, having in mind a better understanding of the structure of these systems, and of the thermodynamic and transport behaviour of this carbohydrate in those aqueous media. On the other words, the main scope of our work shown here is to show the effect of lactose on the diffusion of metal ions involved in biological systems.

In: Lactose
Editors: David Green and Emma Lee
ISBN: 978-1-62808-198-5

Chapter 1

Microbial Transformation of Lactose: Potential of β-Galactosidases for Probiotic and Prebiotic Purposes

Gabriela Zárate[1,2*], *Gabriel Saez*[2] *and Adriana Pérez Chaia*[1]
[1]Centro de Referencia para Lactobacilos (CERELA-CONICET), San Miguel de Tucumán, Argentina
[2]Universidad de San Pablo-Tucumán (USP-T), Argentina

Summary

In the last decades, the interest of food industry has focused in the development of functional foods because of the high demand of healthy foods by consumers around the world. In this sense, foods containing probiotic microorganisms have received particular attention. One of the firsts beneficial properties ascribed to probiotics is their ability to contribute to the intestinal metabolism of lactose from the diet and the amelioration of intolerance symptoms. This effect is mediated by the enzyme β-D-galactosidase of the microorganisms included in the product that can hydrolyze the disaccharide

[*] E-mail: gzarate@cerela.org.ar.

into the food before consumption, as well as to contribute to the intraintestinal digestion of lactose.

However, β-galactosidases have more biotechnological potentialities with a wider spectrum of application. This enzyme possess, depending on the reaction conditions (substrate concentration, aw, pH, reaction temperature), hydrolase or transferase activities, owing the last one to the production of oligosaccharides containing 2 to 10 monomers obtained by transgalactosylation reactions that transfer galactose hydroxyl groups to the disaccharide lactose and/or the different oligomers that are being generated.

Galactooligosaccharides (GOS) are well known prebiotics that modulate the colonic microbiota in a healthy manner and also have other promising industrial applications such as their addition to foods as no-cariogenic, low-calories sweeteners. The raw material used for GOS production could be the lactose from milk as well as the whey discarded by dairy industries from cheeses manufactures. Since whey is considered an environmental contaminant its utilization in this regard seems to be an attractive option.

In the present chapter, the potential of well-known probiotics (such as lactic acid bacteria, bifidobacteria, propionibacteria and yeasts) to contribute to the intestinal metabolism of lactose and/or to produce prebiotic GOS by their β-galactosidase activities is reviewed. In many cases the main biochemical properties of the enzyme of the microorganisms with the higher hydrolytic activities have been determined as well as their abilities to improve lactose digestion in animal models and clinical trials. In this regard, multiple therapeutic strategies have been assessed with different results.

On the other side, the production of GOS by transgalactosylation using *β*-galactosidases from GRAS (Generally Recognized As Safe) microorganisms has also been studied. Depending on the enzyme source and the reaction conditions, the extent of oligosaccharide formation could be as high as 40% of the total sugar content in the solution and three to eleven types of oligosaccharides could be found, including disaccharides other than lactose, trisaccharides and various tetrasaccharides. Although some suitable sources of β-galactosidases that can be used to synthesize novel GOS have been characterized, further studies on the prebiotic effects of these GOS in pure cultures of bifidobacteria, and colonic microbiota of animals and human volunteers are still missing.

Besides, the selection of probiotic microorganisms with appropriate β-galactosidase activities should be complemented with in vivo assays (survival and activity in the intestinal conditions) and certain technological parameters in order to be included in the formulation of novel functional foods, particularly fermented products enriched in GOS and β-galactosidase.

Introduction

Lactose: Some Basic Characteristics

Lactose is a disaccharide made by two simple sugars: glucose and galactose units. It is also known as 4-O-ß-D-galactosylpyranosyl-α-D-glucopyranoside, or in common words as "the sugar of milk". In fact, lactose is the major carbohydrate that occurs naturally in mammal's milk. It is synthesized only in the mammary glands from glucose of blood by joining the two mentioned monosaccharides with a β1,4 glycosidic bond. In non-ruminants animals, most glucose used as substrate for lactose synthesis comes from the digestive tract (derived from the digestion of starch and sugars) whereas in ruminants, much of carbohydrates from silages and grains are fermented by rumen microbes to volatile fatty acids including acetic, propionic, butyric and lactic acids and then resynthesized as glucose in the liver.

Glucose is first converted in the cytosol of mammalian cells to UDP-glucose, which is then converted to UDP-galactose by the UDP-galactose-4-epimerase (GALE). One UDP-galactose and a second glucose molecule are then transported, by the UDP-galactose translocator and the GLUT-1 transporter respectively, into the Golgi apparatus where they become substrates for the lactose synthase enzyme complex, comprised of the enzymatic subunit, galactosyltransferase and its regulatory subunit, α-lactalbumin. Lactose synthase transfers galactose from the donor, UDP-galactose, to glucose and creates the glycosidic linkage of lactose by covalently bonding the monosaccharides. Although galactosyltransferase is found in different tissues in the body, α-lactalbumin is only found in the mammary glands, limiting lactose production to the mammaries.

Lactose concentration varies among the different species of mammals. Human milk contains 70 g/L of lactose approximately whereas bovine, ovine, goat and camelid milks contain 40 to 50 g/L and it is almost undetectable in marine mammals. In milk, lactose exists in two isomeric forms, called α- and β- lactose that differ in the orientation of hydrogen- and hydroxyl groups on C1 in the glucose moiety. Both forms change into another continuously depending on factors like temperature, concentration and pH of the solution. At room temperature the equilibrium results in a ratio of about 40% α-lactose and 60% β-lactose. The existence of these two forms of lactose with different molecular structure greatly influences other properties such as crystallization behavior, crystal morphology, and solubility.

From a technological point of view, lactose is a hygroscopic sugar with a strong tendency to absorb flavors and odors and may cause defects in refrigerated foods such as crystallization, development of sandy texture, and deposit formation (Panesar et al., 2010). On the other side, lactose has a clean sweet taste, without any aftertaste, has low sweetness and is much less cariogenic than sucrose. Furthermore, it represents approximately 30% of the caloric content of cow's milk which is the favorite among occidental culture.

Lactose Metabolism in Mammals

In order to provide energy to the body, lactose has to be hydrolyzed in the small bowel by the specific enzyme known as lactase, lactase phlorizin hidrolase or LPH (EC: 3.2.1.23/3.2.1.62). This enzyme, one of the five intestinal disaccharidases that metabolize carbohydrates of the diet, is localized in the brush border of intestinal villi being its activity particularly high in the enterocytes of jejunum and low or even absent in the proximal duodenum and terminal ileum. As result, the monosaccharides glucose and galactose are released into the intestinal lumen and are actively absorbed into the bloodstream (Gilat, 1972; Semenza et al., 1999)

LPH has two catalytic activities: lactase (β-D-galactoside galacto-hydrolase [EC 3.2.1.23]), which is responsible for the hydrolysis of lactose, and phlorizin hydrolase (glycosyl-N-acylsphingosine glucohydrolase [EC 3.2.1.62]) that hydrolyses β-glycosides with large hydrophobic alkyl chains like those in galactosyl- and glycosyl- β-ceramides.

Cell and molecular biology studies have elucidated the gene structure and transcription, the biosynthesis, enzymatic activity and regulation of lactase protein (Boll et al., 1991, Noren and Sjostriim, 2001). The lactase-activity was first described and characterized by Dahlqvist (1961) and Doell and Kretchmer (1962) and its presence in mammals has been extensively reported (Semenza, et al., 1999).

The gene that encodes for human LPH is located on chromosome 2q21. It covers approximately 55 kb and comprises 17 exons with a length between 79 and 1551 bp, giving rise to a messenger RNA (mRNA) of slightly more than 6 kb. From initiation codon to stop codon, human *LPH* mRNA encodes 1927 amino acids forming the complete translation product (Grand et al., 2003).

The LPH enzyme complex is synthesized in a pre-pro-form as a single-chain precursor (Mr = aprox 220 kDa) which is heavily glycosiylated and proteolytic processed intracellularly to give the mature LPH (Mr= aprox 160 kDa) that reaches the microvillus membrane (Danielsen et al., 1984; Buller et al., 1987; Naim et al., 1987). The primary structure of LPH, deduced from rabbit and human cDNA sequences (Mantei et al., 1988) comprises five domains: (1) a cleavable signal peptide sequence of 19 amino acids at the NH_2 terminus, (2) a large pro-region of 849 amino acids, none of which are found in mature LPH in the brush border membrane, (3) an extracellular domain of 1,014 amino acids, carrying both active sites of the enzymes (lactase and phlorizin hydrolase), (4) a stretch of 19 hydrophobic amino acids serving as a membrane anchor located near the COOH terminus of the protein, and (5) a short cytoplasmic segment of 26 hydrophilic amino acids at the carboxy terminus itself.

Once mature, LPH exerts its function in the outer surface of the brush border by an extracellularly dimer built up of two identical 160 kDa polypeptide chains anchored in the cell membrane.

The enzyme has an optimum pH of 5.8-6.0 and it is inhibited by low pH, glucose, fructose, and mainly by galactose which competes with lactose for reaching the active site. Its rate of hydrolysis is low compared with other disaccharidases and this represents the rate-limiting step in the digestion and absorption of lactose in the intestine.

In all mammals, lactase activity is maximal at the time of birth and during suckling but progressively declines after weaning to approximately 10% of the neonatal value (Wang et al., 1998).

Lactose Metabolism in Microorganisms

Two systems for transport and degradation of sugars inside the cell are known in microorganisms: a) the phosphoenolpyruvate carbohydrate: phosphotransferase system (PEP-PTS) and b) the specific permease system.

The transport and metabolism of lactose has been extensively studied in *E. coli* and lactic acid bacteria but scarcely in other GRAS bacterial genera like *Bifidobacterium* and *Propionibacterium.*

The lactose PEP-PTS is present in strains of lactococci and in many mesophilic lactobacilli (Thompson, 1979, Chassy and Alpert, 1989). In these microorganisms, lactose is phosphorilated during transport through the membrane and enters into the cytoplasm as lactose-6-phosphate being then hydrolyzed by the enzyme phospho-β-galactosidase into glucose and galactose-6-phosphate. The glucose is phosphorilated by a glucokinase and metabolized by the Embden-Meyerhof-Parnas pathway whereas the galactose-6-phosphate is metabolized by the tagatose-6-phosphate pathway. The lactose-PEP–PTS and the enzyme P-β-gal are generally induced by the growth in lactose and repressed by glucose (Kandler, 1983).

In termophilic bacteria of the genus *Lactobacillus*, *Bifidobacterium* (Krzewinski et al., 1996), *Leuconostoc* and *E. coli*, the most frequent system for lactose uptake is the lactose permease. In this case, lactose enters into the cell transported by a specific permease in symport with protons and is hydrolyzed in the cytoplasm by the enzyme β-galactosidase to glucose and galactose. Glucose is immediately metabolized by glucolysis whereas galactose is converted to glucose-6-phosphate by the Leloir pathway entering next to the glycolitic pathway. This system is also inducible but less efficient than lactose PEP-PTS.

In the genus *Propionibacterium*, microorganisms widely used in cheese industry only β-galactosidase activity has been described (Hartley and Vedamuthu, 1975, Zárate et al., 2000) and these observations were confirmed by the identification of the lac operon in the genome of *Propionibacterium freudenreichii* (Falentin et al., 2010).

A singular case that should be mentioned is that of *Streptococcus thermophilus* and *Lactobacillus delbruecki* subsp *lactis* and *bulgaricus*. In these microorganisms lactose enters into the cell in symport with protons by the action of a specific permease and then is hydrolyzed by a β-galactosidase. However, only glucose is metabolized whereas galactose is excreted to the medium by a lactose-H+/galactose antiport system (Hutckins y Poone, 1991).

Figure 1 summarizes main microbial systems for lactose metabolism.

The simultaneous presence of PEP-PTS and the specific permease system for the metabolism of lactose in microorganisms is still discussed as many authors have informed the presence of both β-D-gal and P- β-D-gal in lactic acid bacteria (Hickey et al., 1986) but other consider that this is an artifact since the synthetic substrate used to detect P-β-D-gal (ONPG-6-P) can also be hydrolyzed by the β-D-gal or the alkaline phosphatase releasing the specific substrate (ONPG) for β-D-gal (Fox et al., 1990).

The genes that encode for the enzymes involved in the transport and hydrolysis of lactose can be localized, according to the species, in the chromosome, in one or different plasmids into the same microorganism (Gasson et al., 1987; Chassy et al., 1978; Kanatani et al., 1992); simultaneously in plasmid and chromosome (Fernandez et al., 1999) or one part in a plasmid and another one in the chromosome (Shimizu-Kadota, 1987).

In the case of fungi, lactose seems to be metabolized only by β-galactosidases which are in general extracellular enzymes that have activity at low and high temperatures (4°C and 50°C respectively) with an optimum acid pH range of 2.5 to 5.4 by difference of neutral β-galactosidases of yeasts and bacteria.

Based on structural features and the catalytic mechanism, β-galactosidases are classified in the glycosyl hydrolase (GH) families GH1, GH2, GH35, and GH42 (www.cazy.org). The GH1 family comprises the phospho-β-galactosidase lacG and thermostable β-glycosidases (BglA) with activity on lactose. The GH2 family comprises the lacZ and lacLM type β-galactosidases of *E. coli* and lactic acid bacteria, which use lactose and β(1→3) or β(1→6) linked galactosides as substrate being highly specific for the β-galactosyl residue (however they accept a wide variety of aglycones). GH35 and GH42 β-galactosidases use a variety of β(1→2), β(1→4), and β(1→6) linked galactosides as substrate and some of them prefer other sustrates over lactose (Hinz et al., 2004). Because of the different substrate specificities, it is common to found β-galactosidases of the GH42 and GH2 families simultaneously in the same organism that enable them to adapt more efficiently to substrate availabilities of the environment (Møller et al., 2001; Gänzle, 2012).

Lactose hydrolysis by β-galactosidases takes place through a mechanism of catalysis, which maintains the stereochemistry of the substrate. The first step of hydrolysis leads to the formation of a covalently linked galactosyl-enzyme intermediate. In the second step, the galactosyl moiety linked to the catalytic site is transferred to water. Enzymatic hydrolysis of lactose is accompanied by galactosyl transfer to other sugars, thereby producing oligosaccharides. Retaining glycosyl hydrolases with such a mechanism of catalysis are generally able to synthesize oligosaccharides. This property has been described for GH1, GH2, and GH42 β-galactosidases (Huber et al., 1976). However, their oligosaccharides yields are low in the presence of 2 to 4% lactose (concentrations naturally found in milk) but increase significantly if the concentration of acceptor molecules other than water are sufficiently high, i.e., when the lactose concentration is increased over 30% (Prenosil et al., 1987; Gänzle, 2012).

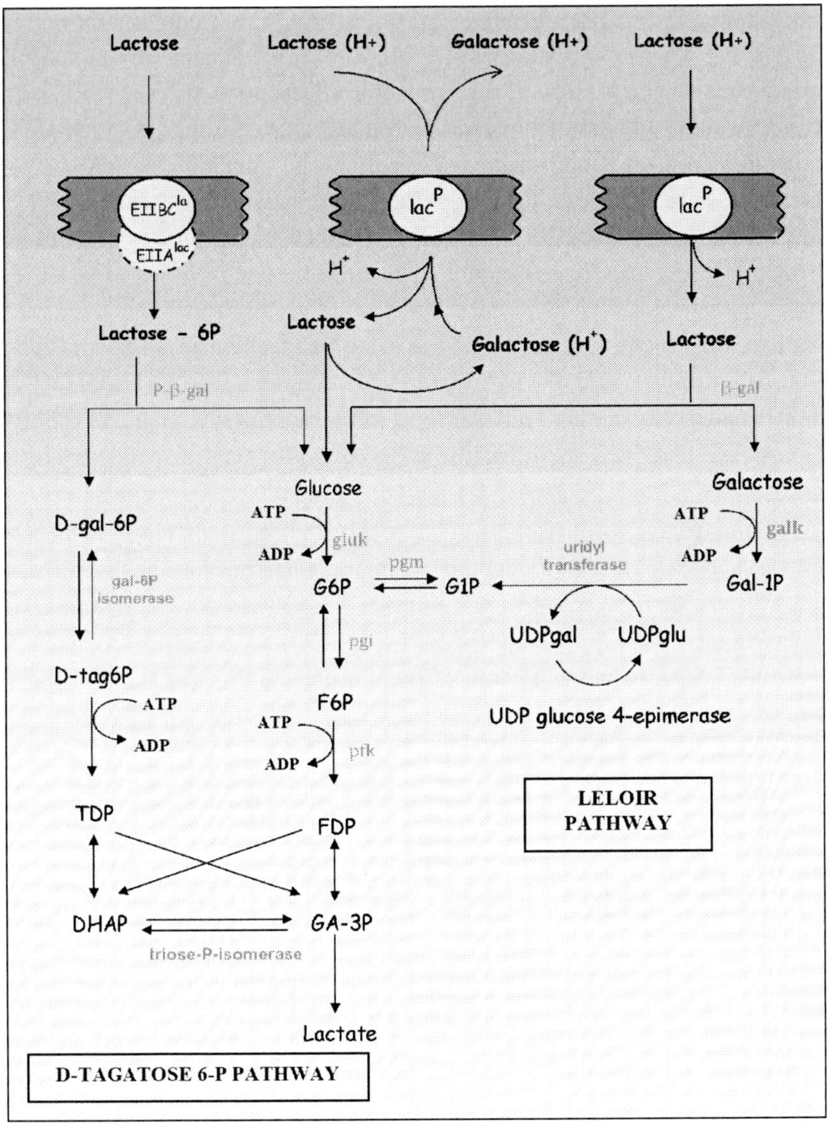

Figure 1. Schematic representation of lactose transport and metabolism in microorganisms.

Lactase Deficiency: Types Known and Causes

Intestinal lactase appears late in fetal development when compared with other disaccharidases. In rats, lactase activity is high during the latter part of gestation and reaches a maximum shortly after birth. In humans, before week 24 of gestation, intestinal lactase activity is low, but begins to increase in the third trimester. It has been estimated that between weeks 35-38 of gestation lactase levels are approximately 70% of the maximal levels observed at birth time and during the suckling period.

Then, it is clear that all young mammals and human infants (except those with a congenital defect) are born with high levels of LPH, which enable them to digest lactose during the first stages of life when milk is the sole nutrient. However, lactase activity declines after weaning in most mammals and in more than half of the world's human population, retaining in adult life a residual activity that represents 5 to 10% of its initial value. By the age of 5 when children are consuming a variety of dairy foods, lactase levels are already low.

In all mammals' species, with the exception of *Homo sapiens,* weaning starts a definitive change towards another type of feeding. Then, an evolutionary law of development is that of discarding everything that won't be need anymore. However, it is important to emphasize that at present, physiological decline of intestinal lactase with age is genetically determined and not an answer to the change of diet of adults (Sahi, 1994; Flatz, 1995). Lactase activity could not be induced by lactose consumption, so that its natural decline cannot be prevented by chronic ingestion of this disaccharide (Semenza et al., 1999).

Since the phenomenon is present in all species, the lost of lactase in adults could be considered a "normal" mammal trait whereas the ability to digest lactose should be the "abnormal" evolutionary innovation.

The adult-onset lactase decline is controlled by a single autosomal recesive gene (Lee and Krasinski, 1998). In fact, the lactase phenotype is explained by the Mendelian system with two alleles: the recessive LACr (lactase restriction gene) and the dominant LACp (lactase persistant gene). As result, homozygous recessive subjects are lactases restricted individuals, those who are heterozygous have intermediate levels of lactase activity and those who are homozygous dominant are lactase persistent subjects.

Other genetically determined mechanisms are also involved in the lactase deficiency in adults, acting both at the post-transcriptional and post-translational levels. In this sense, it has been observed that the characteristics of the LPH-RNAm does not differ between hypolactasic and lactase persistent subjects. The decline in intestinal lactase could be also related to deficient processing and/or homing of an enzymatically inactive precursor from the site of synthesis to the final target membrane. In this respect, it has been reported that intracellular transport and maturation of lactase is retarded in LPH-restricted individuals, leading to an accumulation of newly synthesized LPH in the Golgi and a failure of the enzyme to reach the microvillus membrane (Rossi et al., 1997).

On the contrary, lactase persistence seems to be caused only by an increased biosynthesis of lactase (transcription of the gene) (Boll et al., 1991; Semenza et al., 1999).

The physiological decline of lactase with age is commonly known as "lactase restriction/non persistence", "adult-type/primary hypolactasia"; "primary lactase deficiency" and "primary lactose maldigestion/ malabsorption" whereas those people that retain high levels of lactase in adult life are known as "lactase persistent" or "lactase positive phenotype".

In human beings "primary hypolactasia" is present in approximately 70-90% of adults worldwide but distribution varies among racial/ethnic groups. It is the dominant phenotype in the populations of Oceania, Southeast Asia, Africa and the Americas; whereas lactase persistence is found in Caucasian North Americans, some northern Africans and Europeans (Flatz, 1995; Alliende, 2007).

In addition to primary hypolactasia, other lactase deficiencies are known such as congenital and secondary hypolactasia but they are much less frequent.

Congenital alactasia is the completely absence of lactase at birth and is an extremely rare condition in which persons that suffer this disorder have to follow a lactose-free diet for their entire lives.

Secondary lactase deficiency is, in general, a temporary condition that occurs at any age as a result of injures to the intestinal mucosa by different conditions such as antibiotic and oncologic treatments, gastrointestinal surgery, allergy to milk proteins, viral, bacterial or parasitic enteritis, inflammatory bowel diseases among many others. However, this condition is reversed when the causative factor disappears and the intestine heals (Marteau et al., 1997).

Physiological Consequences of Hypolactasia and Diagnosis of Lactose Intolerance

The non-persistence or deficiency of intestinal lactase causes inevitably an incomplete digestion of lactose after the consumption of milk. In these individuals, the lactose that is not hidrolyzed in the jejunum provokes an osmotic overload in the small bowel that induces water and sodium secretion into the intestinal lumen. This increases the chyme volume and the peristaltic movements of the gut, which in turn, accelerates the transit of intestinal contents (Rambaud et al., 1994). When non digested lactose arrives to the large bowel, it is fermented by the colonic microflora to lactate and *short-chain fatty acids (SCFA)* such as acetate, propionate and butyrate, and gases *like hydrogen, methane and carbon dioxide.* SCFA are rapidly absorbed and used as metabolic fuel by the intestinal cells, and also induce sodium and water recovery. On the other hand, lactose not fermented, unabsorbed volatile acids and gases are responsible of the characteristic symptoms of *lactose intolerance:* abdominal distension, borborygmi, bloating, flatulence, colic, watery diarrhea and acidic stools (Marteau et al., 1997).

Although, incomplete digestion of lactose can lead *to lactose intolerance,* i.e., gastrointestinal discomfort after consumption of milk or its derivatives, not all individuals who do not digest lactose present these unpleasant symptoms. Besides, the degree of lactose tolerance depends on different factors related or not with the host such as the levels of remnant lactase, the speed of gastric empting and intestinal transit; ability of colonic mucosa to absorb gases and volatile fatty acids, composition of the microflora that ferment the non hydrolyzed lactose, amount of lactose ingested, type of dairy product, simultaneous ingestion of lactose with other foods, etc. (Shaw, 1999).

However, it deserves to be emphasized that persons with lactose maldigestion/intolerance consume usually less milk and dairy products than lactase-persistent subjects, having in consequence lower intakes of important nutrients such as calcium, vitamin D, riboflavin, potassium, zinc, phosphorus and magnesium. The inadequate intake of minerals and vitamins affects in the long term their nutritional status and increases the risk of osteoporosis, hypertension, and even colon cancer (Savaiano and Levitt, 2000).

There are several direct and indirect methods to diagnose lactose deficiency/intolerance (Arola, 1994; Shaw, 1999; Vonk et al., 2000; Montalto et al., 2006). First of all, a good medical report that relates the presence of symptoms with the consumption of milk and dairy products is guiding and leads to more specific tests. Quantitative tests may include: the fecal pH determination; the presence of reducing sugars in stools, blood glucose response after an overload of lactose in diet (known as "lactose tolerance test"), the urinary galactose apperarence, the assessment of enzyme activity on biopsy samples of intestinal mucosa and the determination of hydrogen in breath after a load of lactose (Alliende, 2007). In recent years, some genetic tests were developed to diagnose adult-type hypolactasia in children and adults (Rassinpera et al., 2004; Nilsson et al., 2004). However, at present, the "breath hydrogen test" is the most reliable and less invasive method for diagnosis of lactose intolerance.

Treatments for Lactose Intolerance: Exogenous Lactase

Several strategies have been proposed in order to alleviate lactose intolerance symptoms and avoid the elimination of dairy products from the diet, as this usually involves severe nutritional consequences for children and elderly. As mentioned, the decrease in the consumption of dairy products due to lactose intolerance has been linked to deficiencies in vitamins and minerals and the risk of osteoporosis, hypertension and some types of intestinal cancer (Savaiano and Levitt, 2000; Di Stefano et al., 2002).

The first option is the consumption of milk with solid foods, which delay gastric emptying and slow down the intestinal transit of lactose (the nature of foods i.e. yoghurt, cheese, chocolate milk, whole or skim milk, fibers, etc., also influence on the gastrointestinal transit speed). This would result in a higher hydrolysis of the substrate because of the longer exposure to the residual endogenous lactase and less undigested lactose that reaches the colon (Solomons et al., 1985; Martini and Savaiano, 1988; Montalto et al., 2006).

The other proposals involve the use of exogenous sources of lactase activity, especially *β-galactosidases (EC 3.2.1.23)* of microbial origin. This enzyme has been described and characterized in different types of organisms like bacteria, yeasts, fungi, plants and animals (Ramana and Dutta, 1978; Itoh et al., 1980; Hemme et al., 1980; Cesca et al., 1984, Huang et al., 1995, O'Connell and Walsh, 2010; Iqbal et al., 2010; Panesar et al., 2010). However, according to the source, the properties and technological potentialities of lactases differ markedly. In this regard, β-galactosidases of animal and vegetable origin have poor commercial value whereas the β-galactosidases obtained from microorganisms have important technological applications.

Among the most remarkable advantages of microbial sources it could be mentioned the ease of handling of the microorganisms, their rapid growth rates and high production yields. Then, there is an extensive list of microorganisms that have been considered as potential sources of the enzyme (Panesar et al., 2010) (Table 1).

Among the strategies that propose the enzyme-replacement therapy with microbial lactases it could be mentioned in first place, the usage of purified β-galactosidases from fungi and yeasts of the genus *Aspergillus* and *Kluyveromyces* respectively (Grozová et al., 2008; O'Connell and Walsh, 2010), which are frequently added in a liquid form to obtain milks with pre-hydrolysed lactose (lactose-free or low-lactose milks) or included in pills, tablets or drops to be consumed before ingestion of lactose-containing products (Rosado et al., 1984; Solomons et al., 1985; Montalto et al., 2006). Although both approaches are effective in reducing H_2 breath excretion and subjective manifestations of discomfort after milk ingestion, solid lactase preparations are less effective than pre-hydrolyzed milk probably due to the enzyme gastric inactivation (Onwulata et al., 1989).

Several techniques have been assessed for the purification of fungal β-galactosidases. For example, different chromatographic methods (molecular exclusion, ionic and hydrophobic interchange) were used to purify β-galactosidases from cell free extracts of *Apergillus niger*, *Fusarium exysporum* var lini; *Beauveria bassiana* and *Rhizomucor* sp. (Panesar et al., 2010). However, the main disadvantages for the extensive use of these fungal β-galactosidases is the high economic cost of pure enzymes, the lack of practicality (milk has to be pre-hydrolyzed several hours before consumption) and the excessive sweetness of lactose-free milks, which is not always well accepted by consumers (Garcia-Garibay and Gomez-Ruiz, 1996) (Rand et al., 1981; Solomons et al., 1985).

Another strategy that has gained importance in the last decades, is the use of *probiotic microorganisms* that may act in the small and large bowels to contribute to the digestion of lactose and the amelioration of intolerance symptoms (Savaiano and Levitt, 2000, Sherman, 2004; Montalto et al., 2006). *Probiotics* are *"live microorganisms which when administered in adequate amounts confer a health benefit to the host"* (FAO-WHO 2002). Both food and pharmaceutical industries have focused in the development of products containing probiotics due to the great demands of functional foods and alternatives to conventional chemotherapy. It must be emphasized that these microorganisms have to be secure and safe for the consumer (Generally

Recognized As Safe or GRAS status) and that their criteria selection have been clearly stated and revised recently (Morelli and Capurso, 2012).

Table 1. Microbial sources of β-galactosidase

Microbial sources of β-galactosidase		
Bacteria and Archaea	**Fungi**	**Yeasts**
• *Alicyclobacillus acidocaldarius* subsp. *rittmannii* • *Arthrobacter psycrolactophilus* • *Bacillus sp: B. circulans, B. coagulans, B. licheniformis, B. megaterum, B. subtilis, B. stearothermophilus* • *Bacteriodes polypragmatus* • *Bifidobacterium sp.: B. animalis, B. angulatum, B. bifidum, B. infantis, B. longum, B.pseudolongum* • *Clostridiuum acetobutylicum,* • *Corynebacterium murisepticum* • *Enterobacter agglomerans, E. cloaceae* • *Escherichia coli* • *Halomonas sp.* • *Klebsiella pneumoniae* • *Lactobacillus sp.: L.acidophilus, L. bulgaricus, L. helveticus, L. kefiranofaciens, L. lactis, L. sporogenes, L. delbrueckii, L. plantarum, L.reuteri.* • *Leuconostoc citrovorum* • *Pediococcus sp: P. acidilacti, P. pentosaceus* • *Propioionibacterium sp.: P. freudenreichii, P. acidipropionici* • *Pseudomonas fluorescens* • *Pseudoalteromonas haloplanktis* • *Pyrococcus furiosus* • *Streptococcus sp.: S. cremoris, S. lactis, S. thermophius* • *Sulfolobus solfataricus* • *Thermoanaerobacter sp.* • *Thermothoga marítima.* • *Thermus rubus, T. aquaticus* • *Trichoderma reesei* • *Xanthomonas campestres*	• *Alternaria sp.: A. alternate, A. palmi* • *Aspergillus sp: A. foelidis, A. fonsecaeus, A. carbonarius, A. oryzae* • *Auerobasidium pullulans* • *Curvularia inaequalis* • *Fusarium monilliforme, F. oxysporum* • *Mucor meihei, M. pusillus* • *Neurospora crassa* • *Penicillum canescens, P. chrysogenum, P. expansum* • *Saccharopolyspora rectivergula* • *Scopulariapsis sp* • *Streptomyces violaceus*	• *Candida pseudotropicalis* • *Saccharomyces sp.: S. anamensis, S. lactis, S. fragilis* • *Kluyveromyces sp.: K. bulgaricus, K. fragilis, K. lactis, K. marxianus* • *Rhodotorula sp.* • *Sirobasidium magnum* • *Sporobolomyces (Bullera) singularis* • *Sterigmatomyces elviae*

Several studies have shown since the 80s, the ability of lactic acid bacteria and bifidobacteria included in dairy products to enhance lactose absorption in people suffering from intolerance symptoms (Kim and Gilliland, 1983; Gilliland and Kim, 1984; Lin et al., 1991; Montes et al., 1995, Jiang et al.,

1996; Mustapha et al., 1997, Savaiano and Levitt, 2000; de Vrese et al., 2001). In fact, the contribution to lactose metabolism was one of the first accepted beneficial properties for probiotics. Two main mechanisms seem to be involved: a) the reduction of lactose content by fermentation of the product prior to consumption and b) intraintestinal hydrolysis of lactose by the bacterial β-galactosidases (Montalto et al., 2006).

Besides, some probiotic microorganisms, such as bifidobacteria and propionibacteria, can compete and replace (at least temporarily) other components of the colonic microbiota that are gas producers, and supply SCFA (products of their metabolism) to the intestinal lumen. These volatile acids that are readily absorbed by the colonic mucosa, contribute to the energy recovery and stimulate reabsorption of water and salts; counteracting osmotic diarrhea produced by undigested sugars (Jiang and Savaiano, 1997; Zárate, 2001; Pérez Chaia and Zárate, 2005; He et al., 2008).

A final approach that should be mentioned is the chronic ingestion of lactose in order to enhance colonic adaptation of resident microbiota. It has been reported that continuous lactose consumption decreases H_2 excretion and the severity of gastrointestinal symptoms (Hertzler and Savaiano, 1996; Briet et al., 1997). Although the majority of subjects continue to maldigest lactose (since intestinal lactase is not inducible) their tolerance to milk without symptoms is increased. This adaptative phenomenon has been explained by changes in gut microbiota and some colonic functions (motility, transit and pH). Hertzler and Savaiano (1996) showed that long term feeding with lactose altered the ability of fecal bacteria to metabolize it, since β-galactosidase activity increased up to 3 times the control value after 10 days of milk consumption. Besides, the lower H_2 excretion could be result of a decreased production due to reduction of colonic pH, enhanced fermentation capacity of bifidobacteria and other lactic acid bacteria, which metabolize lactose without hydrogen production or increased hydrogen consumption by other colonic bacteria (Perman et al., 1982).

In a study conducted with Japanese lactose malabsorbers, Ito and Kimura (1993) observed that consumption of 15 g lactose/day increased the amount of lactobacilli, enterococci, and SCFA and decreased clostridia and bacteroides in feces within 6 d. It seems that the continuous supply of lactose may modify intestinal microbial populations in a beneficial way by decreasing bacterial metabolic products that cause the adverse symptoms of intolerance.

Some Antecedents on the Potential of Probiotics to Enhance Lactose Digestion

It is well-known that non-fermented and fermented milks with strains of lactic acid bacteria such as *Streptococcus thermophilus*, *Lactobacillus delbruecki* subsp *bulgaricus*, *Lactobacillus acidophilus* and bifidobacteria (*Bifidobacterium bifidum, B.longum, B.infantis; .adolescentis; B. animalis*) which can be either from dairy or intestinal origin improve lactose digestion and ameliorate intolerance symptoms. The beneficial effect of these organisms is attributed mainly to the action of their β-galactosidase enzyme which can survive the passage through the gastrointestinal tract to hydrolyze lactose *in vivo* in the small intestine (Kim and Gilliland, 1983; Gilliland and Kim, 1984; Kolars et al., 1984; Savaiano et al., 1984; Martini et al., 1987a; Lin et al., 1991; Jiang et al., 1996; Mustapha et al., 1997, He et al., 2004; Pérez Chaia and Zárate, 2005; He et al., 2008). Besides, the probiotic bacteria may act by preventing symptoms of intolerance in the large intestine in addition to or rather than improving lactose digestion in the small intestine.

Different factors, such as acid tolerance; bile and pancreatic enzymes, and the gastrointestinal transit rate could influence the delivery and activity of β-galactosidase into the intestine and must be considered in the selection of strains for improving lactose digestion in humans.

Several studies have demonstrated the contribution of fermented products, mainly yogurt, to lactose metabolism in lactose maldigesters. By 1974, Gallagher et al. first observed that fermented milk products were better tolerated than non fermented ones. Later studies determined that lactose in yogurt was better digested than lactose in milk or solution, that lactase activity in the small bowel was increased after yogurt consumption and that viable microorganisms improved digestion better than heat-killed bacteria (Kolars et al., 1984; Gilliland and Kim, 1984; Savaiano et al., 1984, Martini et al., 1987a; He et al., 2004). These effects were explained by the high β-galactosidase activity of bacteria used in the yogurt production, partial hydrolysis of lactose during fermentation (aprox. 20%), protection of microbial cells by buffering capacity and delayed transit time due to the viscous texture of the product.

The better digestion of lactose from yogurt has been observed both in primary and secondary hypolactasic subjects, in infants with diarrhea and elderly and in patients with inflammatory bowel syndromes (Varela-Moreiras et al., 1992; Boudraa et al., 1990; Arrigoni et al., 1994). Furthermore, different strains of lactic acid bacteria used in yogurt (that differ in their β-galactosidase activities) exhibited the same improvement in lactose digestion (Martini et al.,

1991). On the contrary, many fermented milks have been found to be without any significant effect on lactose digestion despite the large variability in their lactase activities (Table 2).

Other lactic acid bacteria such as *Lactobacillus acidophilus* as well as bifidobacteria (mainly *B. longum*) are commonly provided in non-fermented dairy products as "acidophilus" and "bifidus" milks. Most of these products contain less lactase than yogurt, are less viscous (Shah et al., 1992) and improve to a lesser extent lactose digestion (McDonough et al., 1987; Lin et al., 1991; Hove et al., 1994; Montes et al., 1995, Jiang et al., 1996; Mustapha et al., 1997; Zhong et al., 2006; He et al., 2008). Although some products seemed to be successful others were completely ineffective (Levri et al., 2005) (Table 2).

Bile tolerance seems to play an important role in differences observed among the products. Lin et al. (1991) and Mc Donough et al. (1987) have proposed that bile-sensitive strains should be the most effective at improving lactose digestion in intolerant subjects, since the intracellular β-galactosidase may be released from the cells by lysis during passage through the gastrointestinal tract. Mc Donough et al. (1987) reported that unfermented milk prepared with disrupted cells of *L. acidophilus* improved lactose digestion more effectively than milk supplemented with intact microorganisms and suggested that lactose cannot easily access the intracellular β-gal during digestion in bile-tolerant strains. In contrast, Martini et al. (1987a) failed to improve lactose absorption in humans with milk supplemented with sonicated cells of yogurt bacteria probably due to a drastic loss of acid stability of the enzyme. In conclusion, they remarked on the importance of both cell wall and membrane in the protection of the enzyme from gastric proteolytic activity and acidity.

Other authors (Jiang et al., 1996; Mustapha et al., 1997) showed that the most bile tolerant strains of *L. acidophilus* and bifidobacteria (although their β-galactosidase activity was low) contributed to a great reduction of breath H_2 excretion and significantly reduced symptoms of lactose intolerance. Noh and Gilliland (1993; 1994) demonstrated that bile did not promote cell lysis or the leakage of the enzyme β-galactosidase from the cells, and instead bile salts increased cellular permeability, permitting more substrate to enter the cells to be hydrolyzed.

In recent years, it has been proposed that the construction of a β-galactosidase food-grade expression system could provide a new solution for the management of lactose intolerance. Zhang et al. (2011) successfully constructed the recombinant strain *Lactococcus lactis* MG1363/FGZW expressing food-grade β-galactosidase, by using pMG36e in which the β-galactosidase gene (lacZ gene) of *L. bulgaricus* wch9901 was inserted. This recombinant *L. lactis* exhibited a higher β-galactosidase activity compared to *L. bulgaricus* wch9901 (Wang et al., 2008). Besides, its efficacy on alleviating lactose intolerance symptoms, such as watery diarrhea, in post-weaning Balb/c mice used as animal model was demonstrated (Li et al., 2012). These results might promote the usage of gene-modified microorganisms and related technology both in health and food fields.

Dairy propionibacteria are microorganisms of interest since they are extensively used for manufacture of Swiss-type cheeses, biological production of propionic acid and vitamin B12 (Langsrud et al., 1973; Quesada Chanto et al., 1994). In recent years, many potential probiotic properties have been reported for these microorganisms (Pérez Chaia et al., 1999, Cousin et al., 2010; Zárate, 2012).

β-galactosidase activity of propionibacteria was first demonstrated by Hartley and Vedamuthu (1975) for strains of *P. shermani*. In the same way, we described and characterized for the first time the β-galactosidase enzyme of *P. acidipropionici* (Zárate et al., 2002). Further studies on regulation of β-gal activity by growth carbon sources, like lactose and lactate, were also developed to decide the most appropriate vehicle to deliver propionibacteria with the highest β-galactosidase to the host (Zárate and Pérez Chaia, 2012). Maximum enzyme activity was observed at pH 7.0 and 50°C, but a high activity was even detected at 37°C. Our results suggest that the environmental conditions commonly present in the human's intestine may be adequate for the synthesis and activity of β-galactosidase from these *Propionibacterium* strains. Besides, the enzyme resists the cooking temperature of Swiss-type cheese and most of the cations present in milk stimulate its activity suggesting that a probiotic cheese would be a suitable vehicle for delivery of β-galactosidase of propionibacteria to the host.

As mentioned previously, it is thought that to digest lactose in vivo in the intestine the selected microorganisms and their β-galactosidase must be able to survive the harsh conditions of gastrointestinal transit. In this respect, we demonstrated that the β-galactosidase of *P. acidipropionici* strains consumed both in skim milk or the aqueous phase of an Emmental cheese remain unaltered or were stimulated in the conditions of the human's gut. Furthermore cells were permeabilized by bile, which in turn may favour the hydrolysis of lactose within the small bowel (Zárate et al., 2000a; 2000b) in agreement with conclusions of Noh and Gilliland (1993, 1994).

Our results suggest that if bile-tolerant strains of *P. acidipropionici* are ingested in a Swiss-type cheese, a large number of cells would survive the passage through the stomach, reach viable to the intestine, and could contribute in situ to the lactose metabolism (Zárate et al., 2000b, Pérez Chaia and Zárate, 2005).

Further studies in an animal model demonstrated that feeding mice with *P.acidipropionici* CRL 1198 included in milk or cheese increased the propionibacteria counts and β-galactosidase levels in the small and large bowels and the propionic acid concentration in the caecum (Figures 2 and Table 3). The increase of SCFA in the colon could have relevance by counteracting the osmotic diarrhea induced by non-digested carbohydrates (Zárate, 2001; Perez Chaia and Zárate, 2005).

Other studies have also intended to modify the composition and metabolic activities of the colonic microbiota in order to improve lactose tolerance by introducing probiotic microorganisms such as lactobacilli and bifidobacteria (Jiang and Savaiano, 1997a; 1997b, He et al., 2008).

Table 3. Bacterial β-galactosidase activity and counts of propionibacteria in intestinal contents of mice fed different diets* (Data reproduced from Table III of Pérez Chaia and Zárate, 2005, Lait 85, 85-98)

	Viable counts ($\log_{10}$ cfu g^{-1})		β-galactosidase activity (μmol h^{-1} g^{-1})	
Diets	**Small bowel content**	**Cecal content**	**Small bowel content**	**Cecal content**
Water	4.9 ± 0.9	5.8 ± 0.8	12.6 ± 2.8	380.6 ± 33.2
Milk	5.8 ± 0.5	6.4 ± 0.3	23.5 ± 3.0a	460.3 ± 62.4a
Milk+bacteria**	8.5 ± 0.9a	9.5 ± 0.7a	32.8 ± 3.8b	520.9 ± 51.5a
Cheese**	8.0 ± 0.6a	8.8 ± 0.3a	40.2 ± 2.7c	518.6 ± 13.8a

* The different groups were fed a solid conventional diet and drank/ate water, milk, milk with propionibacteria or cheese. Viable counts of propionibacteria (log10 cfu·g–1) and β-galactosidase activity were determined in the mice intestinal contents at the end of the 7 d treatments. Results are means ± sd (n = 5).

Different superscripts in values of the same sample indicate significant differences between treatments ($P < 0.05$).

** Bacterial dose was adjusted to provide 10^9 cells of *P. acidipropionici* CRL1198 per day.

Table 2. References of some studies that have positive and negative results on improving lactose maldigestion by using probiotics in animal models or clinical trials

Some positive results			
Reference	**Product**	**Microorganisms**	**Main result**
1. Gallagher et al (1974)	Yogurt	Starter bacteria: *S. thermophilus, L.bulgaricus*	Better tolerance than milk
2. Kim & Gillliland (1983)	Acidophilus milk	*L.acidophilus*	↓ breath H_2
3. Gilliland & Kim (1984)	Yogurt	Starter bacteria	↓ breath H_2
4. Kolars et al (1984)	Yogurt	Starter bacteria	↓ breathH_2, ↓ symptoms
5. Savaiano et al (1984)	Yogurt	Starter bacteria	↓ breath H_2
6. Martini et al (1987b)	Yogurt	Starter bacteria	↓ breath H_2
7. Boudraa et al. (1990)	Yogurt	Starter bacteria	↑ symptoms in children
8. Lin et al. (1991)	Yogurt	Starter bacteria	↓ breath H_2
9. Martini et al. (1991)	Yogurt or milk+ probiotics	SB + *L. acidophilus* or *B.bifidus*	↓ breath H_2
10. Lin (1991)	Acidophilus milks	*L.acidophilus*	↓ breath H_2
11. V-Moreiras et al. (1992)	Yogurt	Starter bacteria	↓ breath H_2 in elderly
12. Arrigoni et al. (1994)	Yogurt	Starter bacteria	↑ tolerance in short bowel syndrome
13. Kotz et al. 1994	High lactase yogurt	Starter bacteria	↓breath H_2
14. Montes et al. (1995)	Acidophilus milks	SB or *L.acidophilus*	↓breath H_2 in children
15. Jiang et al. (1996);	Milk+bifidobacteria	*Bifidobacterium longum*	↓breath H2 when grown on lactose
16. Mustapha et al. 1997,	Acidophilus milks	Strains of *L.acidophilus*	Some strains ↓breath H2
17. P. Chaia & Zárate 2005;	Milk+propionibacteria	*P. acidipropionici*	↑ β-gal and SCFA
18. Zhang et al. (2006)	Yogurt+bifidobacteria	SB+ *B.animalis* or capsules of *B.longum*	↑ symptoms of intolerance
19. He et al. 2008	Yogurt+bifidobacteria	SB+ *B.animalis* or capsules of *B.longum*	↑ tolerance of intolerance
Some negative results			
20. Newcomer et al. (1983)	Acidophilus milk	*L. acidophilus*	No improve in tolerance symptoms
21. Dehkordi et al. (1995)	Acidophilus milk	*L. acidophilus*	No decrease of H_2
22. Vesa et al. (1996)	Fermented milk	*L. acidophilus* and bifidobacteria	No decrease of H_2
23. Saltzman et al. (1999)	Vials of bacteria	*L. acidophilus*	No decrease of H_2 and symptoms

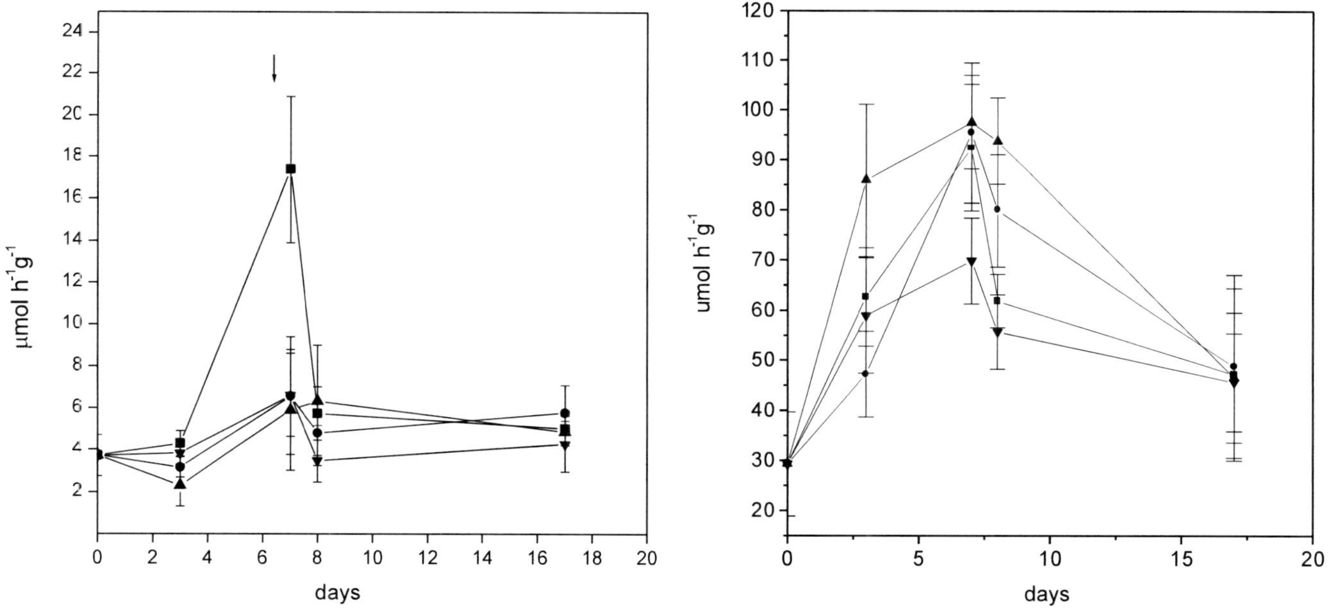

Figure 2. β-galactosidase activity of small bowel (left) and cecal contents (right) of mice fed with milk or different strains of dairy propionibacteria. Symbols: (▼) milk; (■) *P.acidipropionici* CRL 1198; (●) *P. freudenreichii* CRL 757; (▲) *P. acidipropionici* Q4. (Zárate, 2001).

In a recent study, He et al. (2008) investigated whether a 2-week supplementation with capsules containing *Bifidobacterium longum* and a yogurt enriched with *B. animalis* could modify the composition and metabolic activities of the colonic microbiota in 11 Chinese lactose-intolerant subjects. The results showed that supplementation with bifidobacteria modified the amount of colonic bacteria (with no change in composition rather than a tendency of bifidobacteria to increase during and after consumption), increased the β-galactosidase activity in faeces and alleviates symptoms in lactose-intolerant subjects. Both bifidobacteria tested remain transiently present during the supplementation period, but disappeared after the consumption stopped as was reported for many other probiotic strains (Tannock et al., 2000; Massi et al 2004; Zárate, 2012). It has been stated that once the adult intestinal microbiota get established, colonization with exogenous strains is usually difficult and transient, and continuous consumption is need for their maintenance (Mackie et al., 1999).

From these previous studies it could be suggested that modulation of colonic metabolism, such as β-galactosidase activity and SCFA production, through dietary supplementations should be another useful approach for the management of lactose intolerance.

Other Biotechnological Applications of Microbial β-Galactosidase Enzyme: Food Industry and Prebiotics

Besides its application in health by means of probiotics, microbial β-galactosidase can be used for the hydrolysis of lactose in milk and whey being in consequence of great interest for food and dairy processing industries.

Treatment of milk and milk products with β-galactosidases to reduce their lactose content is considered an adequate method to increase their potential uses and solve the problems of lactose insolubility, crystallization and lack of sweetness. Hydrolysis of lactose is recommended for the production of condensed milk, ice cream, low-lactose products, dried whey and animal nutrition. Moreover, treated milks become a more suitable food for intolerant people as previously mentioned whereas the hydrolysis of lactose in whey converts it into sweet syrup, which can be used in various processes of dairy, bakery and soft drink industries (Panesar et al., 2010).

For industrial hydrolytic purposes the enzyme can be used in either soluble or immobilized forms. The free enzyme or soluble form can be used only for batch processes whereas immobilization allows batch and continuous operation. Furthermore it makes possible the reutilization of the enzyme without the need to separate the cells from the milk/whey following processing and also improve the enzyme stability.

Grozová et al. (2008) and Panesar et al. (2010) have reviewed extensively the different methods used for purification and immobilization of microbial β-galactosidases, their efficiencies in hydrolyzing lactose and their potential applications in food industry. The enzyme has been immobilized by various methods such as physical adsorption, entrapment, and covalent binding on different supports (Mateo et al., 2007). Enzyme adsorption is the simplest method of immobilization and uses different inorganic (alumina, silica, porous glass, ceramics, diatomaceous earth, clay, bentonite, etc.) and organic (cellulose, starch, activated carbon and ionexchange resins, such as Amberlite, Sephadex, Dowex) support materials. Matrix and membrane entrapment (including microcapsulation) are the major methods of entrapment. The matrices used for the immobilization include polymeric materials such as Ca-alginate, agar, *k*-carragenin, polyacrylamide and collagen as well as solid matrices like activated carbon, porous ceramic, and diatomaceous earth. The membranes commonly used for the entrapment of enzymes are nylon, cellulose, polysulfone, polypropylene and polyacrylmide. Immobilized enzymes may be retained on support surfaces by covalent binding. The three methods have been applied mainly to species of *Apergillus, Kluyveromyces*, and bacteria such as *E. coli*, *Bacillus circulans, B. stearothermophilus, S. thermophilus, L. bulgaricus* and *Thermus aquaticus*. However, some problems associated with the immobilized enzyme system may be present such as microbial contamination, protein adherence, and channeling. The periodic washing and pasteurization and flow direction of feed can partially solve these problems (Panesar et al., 2010).

At present, immobilized β-galactosidases by the mentioned methods are being used for lactose hydrolysis of milk, whey/whey permeate and for the production of galactooligosaccharides (Haider and Husain, 2009; Gosling et al., 2009, 2010; Klein et al., 2013). Milk with hydrolyzed lactose may be used for the manufacture of yogurt and cheeses. From a technological point of view, hydrolysis of lactose prevents lactose crystallization and texture defects and accelerates the development of flavor and structure of these products. In the case of lactose of whey, concentrated hydrolyzed whey or whey permeates can be used as sweeteners in canned fruit syrups and soft drinks. Whey treatment also contributes to solve a major environmental problem since its disposal in water streams significantly increases their biological oxygen demand.

Optimum pH and temperature for the hidrolytic activity vary according to the microorganism used as source of the enzyme. For example, optimal pH values ranging from 2.5 to 4.5 have been reported for filamentous fungi, whereas pH values ranging from 6.0 to 7.0 have been reported for yeasts and bacterial β-galactosidases. Unlike pH, the optimal temperature values are found between 35 and 50°C, regardless the origin of the enzyme, although activities below 10°C and above 70°C have been reported. Accordingly, acid-pH enzymes from fungi are used for hydrolysis of acid whey and whey permeate whereas the neutral-pH enzymes from yeasts and bacteria are appropriate for processing of milk and sweet whey (Panesar et al., 2010). The immobilized enzyme preparations have shown to be successful in hydrolyzing up to 99% of lactose present in milk or whey. Besides, both pyschrophilic and termophilic bacteria have advantages for processing products at low temperatures and for retaining activity at higher temperatures during prolonged periods preventing contaminations (Pessela et al., 2003, Nakagawa et al., 2006, Wang et al., 2013).

As example of the described application it could be mentioned that Rhimi et al. (2010), purified and biochemically characterized the recombinant β-galactosidase from dairy *S. thermophilus* LMD9 strain showing that immobilization in alginate beads enhanced stability of the enzyme and activity at various temperatures, including 4 and 50°C, and at pH values from 4.0 to 8.5. The remarkable bioconversion rates of lactose in milk and whey at different conditions evidenced the efficiency of this enzyme as biocatalyst and promote its use for lactose hydrolysis in milk and other dairy products.

A final approach of exploiting β-galactosidase that deserves attention is the use of the enzyme for the *synthesis of galactooligosaccharides (GOS).*

In addition to its hydrolyzing ability, β-galactosidase enzyme can catalyze transgalactosylation reactions, where glycosyl group of one or more D-galactosyl units is transferred onto the D-galactose moiety of lactose resulting in oligosaccharides with a higher degree of polymerization (Otieno, 2010).

These molecules, formed by galactose transfer, containing one or more galactosides are named *galactooligosaccharides or GOS.*

Galactooligosaccharides are also defined as nondigestible oligosaccharides composed by 1 molecule of glucose and a variable number of galactose units (usually 2 to 10 molecules) linked mostly by β(1–4) and β(1–6) glycosidic bonds.

It must be emphasized that lactose hydrolysis and transgalactosylation catalyzed by β-galactosidase are simultaneous reactions in which the enzyme hydrolyzes lactose and, instead of transferring the galactose unit to the hydroxyl group of water, it transfers galactose to another carbohydrate, in this case lactose, resulting in monomeric products, as well as newly formed β-glycosides, mainly di-, tri-, and tetrasaccharides (Prenosil et al., 1987). GOS formed also serve as substrates for both hydrolysis and as acceptors for galactosyl transfer; therefore complex product mixtures may result. Because of the concomitant hydrolysis of glycosidic linkage and synthesis, the yield of a GOS reaction will depend on the ratio of the transferase activity to the hydrolytic activity of the enzyme. However, it has been reported that the equilibrium tends to the direction of hydrolysis leading in general to low GOS yields (Hansson et al., 2001a).

GOS have been regarded as "prebiotics", a special class of components of functional foods. *Prebiotics* have been defined as "selectively fermented ingredients that allow specific changes, both in the composition and/or activity of the gastrointestinal microflora which confers benefits upon host well-being and health" (Roberfroid, 2007). In practical words, it has been stated that prebiotics are not digested in the small bowels of humans and animals but stimulates the growth of desirable bacteria in the colon leading to recognized health benefits (Macfarlane et al., 2008).

Oligosaccharides such as inuline, fructooligosaccharides and galactooligosaccharides have been particularly proposed as useful dietary tools for the modulation of the colonic microbiota toward a healthy balance, which usually involves the selectively increasing of bifidobacteria and lactobacilli at the expense of less desirable organisms such as clostridia, bacteroides and enterobacteria (Fuller and Gibson, 1998). Furthermore, other health benefits have been ascribed to GOS such as prevention of diarrhea and constipation, inhibition of the attachment of pathogenic bacteria to the colonic epithelium, prevention of bone loss, reduction of serum cholesterol, reduction of colon cancer risk and enhancement of immunity (Sako et al., 1999; Vulevic et al., 2008; Sinclair et al., 2009). Besides effects on health, GOS have other useful properties: their stability in acidic conditions and during food processing (pasteurization and baking temperatures) make them useful ingredients for a wide variety of products such as bread, jams, confectionery, acid beverages and fermented milk products (Sako et al., 1999). Because of their taste quality and low sweetness GOS are considered functional sweeteners: they have low cariogenicity and low caloric value as they cannot be metabolized by microorganisms from the oral cavity neither absorbed in the small bowel (Sako et al., 1999; Mac Farlane et al., 2008).

A recent study, demonstrated the capacity of galactooligosaccharides to protect particularly sensitive strains to preservation processes like freeze–drying. This novel role of GOS could be useful in the formulation of self-protected synbiotics in which probiotics would be combined with galactooligosaccharides that exert both a prebiotic and protecting effect (Tymczyszyn et al., 2011).

In view of all these properties and the high demand of healthy foods by consumers, there is a growing commercial interest of food industry for producing GOS as functional ingredients. As result, many researches have been developed in the last years (Barreteau et al., 2006; Mlichova and Rosenberg, 2006; Trincone and Giordano, 2006; Macfarlane et al., 2008, Panesar et al., 2010; Gosling et al., 2010; Otieno, 2010; Ganzle, 2012). Furthermore, at least 47 patents regarding β-galactosidases and GOS production and applications have been published since 2000. Patents include the production of GOS with native or recombinant enzymes, the production of β-gal resistant to low or high temperatures and the development of functional foods and cosmetics (Bayfield et al., 2010).

Although β-galactosidase catalyzes both hydrolysis and transgalacto-sylation reactions, the process conditions for lactose hydrolysis and GOS synthesis are different.

Factors influencing GOS production and yield may include enzyme origin and reaction conditions such as: pH, temperature, aw, substrate concentration, length of reaction and process type. As a general rule, transgalactosylation is favored (more and larger GOS are obtained) at elevated temperature, high lactose concentration and low water activity in the reaction medium (Otieno, 2010; Gosling et al., 2010). The amounts and composition of the GOS, as well as the linkages types between the galactose units also depend on the aforementioned factors (Prenosil et al., 1987, Mahoney, 1998, Splechtna et al., 2007; Gänzle, 2012).

In most cases, depending on enzyme sources and conditions, the yields of GOS could be as high as 40-42% and three to eleven types of oligosaccharides could be found, being disaccharides other than lactose, trisaccharides and various tetrasaccharides the dominant forms. These yields could represent the upper limit for what is possible to achieve by wild-type β-glycosidases (Onishi et al., 1995). In a recent study, Park et al. (2008) observed that the β-galactosidase from *Sulfolobus solfataricus* produced under the conditions of pH 6.0, 80°C, 600 g lactose L^{-1}, and 3.6 U enzyme mL^{-1}, 315 g GOS L^{-1} with a yield of 52.5% (w/w). This represents one of the highest concentrations and yields of GOS produced by thermostable β-galactosidases reported to date. However, production of GOS by the β-galactosidase of this microorganism was even further increased by site-directed mutagenesis allowing GOS yields to reach 61.7% (Wu et al., 2013).

As mentioned above, β-galactosidase from different microorganisms yields varying concentrations of oligosaccharides with different saccharide compositions and degrees of polymerization. For example, Hsu et al. (2007), showed that β-galactosidase from *Bifidobacterium longum* BCRC 15708 produced 2 types of GOS, tri and tetrasaccharides, from a lactose concentration of 40% being trisaccharides the major type of GOS formed; whereas *B. bifidum* NCIMB 41171 gave rise to a variety of oligosaccharides with different degrees of polymerization (DP > 3) and transgalactosylated disaccharides from lactose and whey permeate (Goulas et al., 2007, Osman et al., 2010).

The enzyme source also influences the linkage type of GOS. For example, β-galactosidases derived from *B. circulans* synthesize GOS with mixed β(1→2), β(1→3), β(1→4) and β(1→6) linkages. The β-galactosidases from *Geobacillus stearothermophilus* and from *Bifidobacterium infantis* produce predominantly β(1→3) linked GOS; *Sterigmatomyces elviae* and *Kluyveromyces lactis* form β(1→4) linkages; whereas β-galactosidases from *Lactobacillus reuteri*, *Streptococcus thermophilus* or *B. bifidum* synthesize GOS with β(1→3), or β(1→6), linkages (Toba et al., 1981; Dumortier et al., 1990; Onishi et al., 1995; Yanahira et al., 1998; Hung et al., 2001; Splechtna et al., 2006; Placier et al., 2009).

The usage of free or immobilized β-galactosidases as well as batch and continuous processes also influence GOS yields. Both immobilization and continuous processes favor transgalactosylation and therefore lead to higher GOS yields (Otieno, 2010). Shin et al. (1998) observed that an immobilized enzyme in a continuously stirred-tank reactor (CSTR) with a lower lactose concentration yielded higher GOS than a batch process with a higher lactose concentration involving the same enzyme obtained from *Bullera singularis*.

Immobilization of β-galactosidases for GOS production has been achieved on a variety of supports without altering the enzyme activity or the preference for transgalactosylation. Compared with the corresponding soluble enzymes, several immobilized β-gal systems exhibited reduced product inhibition by glucose, and a higher stability at denaturing temperatures (Petzelbauer et al., 2002; Neri et al., 2009).

In the last years, studies have focused on the screening of sources, purification and characterization of new microbial β-galactosidases with high hydrolase and transferase activities, in order to obtain GOS for industrial and prebiotics purposes. Further strategies have been used to increase the yield of GOS such as reconstruction of the enzyme by protein engineering (Jorgensen et al., 2001; Hansson et al., 2001b; Placier et al., 2009; Iqbal et al., 2010; Wu et al., 2013) and optimization of transglycosylation conditions (Hanson and Adlercreutz, 2001; Hsu et al., 2005; Gosling et al., 2010, Osman et al., 2010).

In this respect, the enzyme has been characterized in mesophilic microorganisms like *Aspergillus oryzae* (Iwasaki et al.,1996), *Bacillus circulans* (Mozzafar et al., 1984; Yanahira et al., 1998), *Bifidobacterium* species: *B angulatum, B.adolescentis, B. bifidum; B. infantis; B.longum, B.pseudolongum* (Møller et al., 2001; Rabiu et al., 2001, Hsu et al., 2005, 2007; Goulas et al., 2007), *Bullera singularis* (Cho et al., 2003), *Escherichia coli* (Cho et al., 2003), *Kluyveromyces lactis* (Chockchaisawasdee et al., 2005), *Lactobacillus* species: *L.reuteri, L.plantarum* (Nguyen et al., 2006; Splechtna et al., 2006; Iqbal et al., 2010); and *Sirobasidium magnum* (Onishi and Tanaka, 1997) among others. However, these microorganisms are unable to produce GOS at a high temperature, a desirable condition for improving GOS production while limiting potential microbial contaminations within the process.

Therefore, GOS production by thermostable β-galactosidases from thermophilic microorganisms such as *Pyrococcus furiosus* (Bruins et al., 2003), *Rhodotorula minuta* (Onishi and Yokozeki, 1996), *Saccharopolyspora rectivirgula* (Nakao et al., 1994), *Sterigmatomyces elviae* (Onishi and Tanaka, 1995), *Sulfolobus solfataricus* (Reuter et al.,1999), *Thermus* sp. Z-1 (Akiyama et al., 2001), and *Thermotoga maritima* (Ji et al., 2005) has been reported. These β-galactosidases have better stability properties that make them an attractive alternative to the thermolabile enzymes for synthesis of GOS at high temperatures.

In recent years, genetic engineering has become a viable option for obtaining higher GOS yields and several homologous and heterologous recombinants have been successfully constructed (Ji et al., 2005; Chen et al., 2008, Iqbal et al., 2010). Some advantages have been ascribed to these recombinant enzymes in comparison with their native counterparts: higher efficiencies, more stability at higher temperatures, ease to purify and large-scale production rather easily achievable (Ji et al., 2005).

As examples it could be mentioned that the thermostable β-galactosidase from *Bacillus stearothermophilus* was expressed in *B. subtilis* WB600 displaying a high level of transgalactosylation activity in the hydrolysis of lactose in milk. The results suggest that this recombinant enzyme may be suitable for both the hydrolysis of lactose and the production of GOS in milk processing (Chen et al., 2008). By their side, Placier et al. (2009) reported a significant increase in the yield of trisaccharides by the mutagenesis of Arg109 to tryptophan in β-galactosidase from *Geobacillus stearothermophilus*.

Recently, a method involving GOS production with a recombinant enzyme was patented. In this case, a modified enzyme from *Bifidobacterium bifidum* was produced by deleting the sequence coding for 580 amino acid residues at the C-terminal of the β-galactosidase gene. The recombinant truncated protein produced in *E.coli* showed higher transgalactosylating activity than the observed with native protein, even at low lactose concentrations, and with a significant reduction in hydrolytic activity (Jørgensen et al., 2001; 2007).

Amongst microbial producers of GOS, *Aspergillus oryzae, A. niger*, *Kluyveromyces lactis,* and *K. fragilis* are commonly referred as the best producers. Some of these microorganisms and the bacteria *Streptococcus thermophilus*, are comercially used for manufacture of GOS available at market such as Oligomate 55 (Yakult, Tokyo, Japan). However, the yeast *Sterigmatomyces elviae* CBS8119 has been found to be the highest producer of GOS among yeasts (Onishi and Tanaka 1995) and the β-galactosidase from *Bacillus circulans* was cited as the best for GOS production among bacterial sources.

Some researches dealing with the enzymatic production of GOS by fungi, yeasts and bacteria are summarized in Table 4.

GOS are primarily synthesized from lactose of milk or dairy products by the direct addition of β-galactosidase. However lactose in aqueous solution and whey from cheese industry can be used as substrate as well. The reutilization of a dairy byproduct such as cheese whey represents an interesting technological strategy that may contribute to solve the environmental problem linked with whey disposal (Audic et al., 2003).

Microbial β-galactosidases can be used as live whole cells carrying the crude enzyme that provides both hydrolytic and transferase activity (Goulas et al., 2007) or the enzymes can be extracted and purified to obtain a stable form with higher activity that leads to a greater GOS yield (Panesar et al., 2010).

Table 4. Galactooligosaccharide (GOS) yield from lactose solutions by different representative microorganisms

Source	Conditions	GOS yield	Reference
Apergillus oryzae	38% initial lactose, 40°C and pH 4.5; batch	31%	Iwasaki et al. (1995)
Apergillus oryzae	50%initial lactose, 40°C and pH 4.5; batch.	52%	Neri et al. (2009)
Apergillus oryzae	40% initial lactose, 40°C and pH 4.5; continuous.	26.6%	Albayrak and Yang (2002)
Penicillium simplicissimum	60% initial lactose, 50°C and pH 6.5, batch	30.5%	Cruz et al. (1999)
Saccharopolyspora rectivirgula	60% initial lactose, 70°C, and pH 6.0 batch	41%	Nakao et al. (1994)
Bullera singularis	20% actose, 50°C and pH 5.0, batch	50%	Cho et al. (2003)
Kluyveromyces lactis	25% initial lactose, 40°C and pH 6.5, batch	17,1%	Martinez- Villaluenga et al. (2008)
Kluyveromyces lactis	40 % initial lactose, 40°C, pH 7.0 , batch and continuous	24,8%	Chockchaisawasdee et al. (2005)
Sterigmatomyces elviae	20% initial lactose, 60°C pH 5.0	39%	Onishi et al. (1995)
Bacillus circulans	4.56% lactose, 40°C and pH 6.0, continuous.	40%	Mozaffar et al. (1986)
Bacillus singularis	10% lactose, 45°C and pH 4.8, continuous.	55%	Shin et al. (1998)
Bifidobacterium adolescentis	5 to 30% initial lactose, 55°C and pH 7.5, batch (shaking).	43.1%	Rabiu et al. (2001)
Bifidobacterium angulatum	5 to 30% initial lactose, 55°C and pH 7.5, batch (shaking).	43.8%	Rabiu et al. (2001)
Bifidobacterium bifidum BB-12	5 to 30% initial lactose, 55°C and pH 7.5, batch (shaking).	37.6%	Rabiu et al. (2001)
Bifidobacterium infantis	5 to 30% initial lactose, 55°C and pH 7.5, batch (shaking).	47.6%	Rabiu et al. (2001)
Bifidobacterium longum	40% initial lactose, 45°C and pH 6.8	32.5%	Hsu et al. (2007)
Bifidobacterium pseudolongum	5 to 30% initial lactose, 55°C and pH 7.5, batch (shaking).	26.8%	Rabiu et al. (2001)
Lactobacillus reuteri	205 g/l 37 6.5 50 mM phosphate containing 1 mM MgCl2 80	38%	Splechtna et al. (2006)
Pyrococcus furiosus	745 mmol/kg lactose, 80°C an pH 5.0	18.1%	Bruins et al. (2003a)
Pyrococcus furiosus	70% initial lactose, 95°C and pH 5.0, batch	45%	Hansson et al. (2001)
Sulfolobus solfataricus	60% inicial lactose, 80°C and pH 6.0	52.5%	Park et al. (2008)
Thermotoga maritima	50% initial lactose, 80°C and pH 6.0	19%	Ji et al. (2005)

The idea of using whole-cell as enzyme sources for GOS synthesis is usually applied to probiotic microorganisms that could provide double advantages by acting alternatively as probiotics or prebiotics factories.

In this regard, the majority of the studies have focused in the production of GOS by β-galactosidases of *Lactobacillus* and *Bifidobacterium* species as these microorganisms are the target of prebiotics in the colon. It was suggested that β-galactosidases from probiotic microorganisms might produce galactooligosaccharide structures that have special prebiotic effects, specifically targeting selected colonic probiotic strains (Rastall and Maitin, 2002). However, to exert prebiotic effects the different glycosidic bonds present in a mixture of oligosaccharides have to be cleaved to different extents by the target colonic microorganisms to be next fermented.

It has been suggested that it is probable that probiotic β-galactosidases, which rapidly hydrolyze certain glycosidic linkages, can preferentially form the same bond types when acting in the transgalactosylation mode. In order to test the ability of probiotics to syntethesize GOS with distinct profiles and specific selectivity, Rabiu et al. (2001) extracted β-galactosidases from different *Bifidobacterium* species and used them in glycosyl transfer reactions to synthesize oligosaccharides from lactose. Distinct product spectra were obtained for each species and these were different to the composition of Oligomate 55, a commercial GOS product obtained with fungal enzyme. Increases in growth rate on homologous galactooligosaccharide (compared with Oligomate 55) were seen for *B. angulatum*, *B. bifidum* BB-12, *B. infantis*, and *B. pseudolongum*. Therefore, the use of bifidobacterial β-galactosidases, instead of other microbial sources, could potentially yield GOS with improved prebiotic selectivity. However it remains to elucidate whether the specific oligosaccharides generated can select for the homologous probiotic in a mixed population like colonic microbiota.

In a study about the relationship between various disaccharides and their prebiotic effect, it was shown that among a group of galactose-containing disaccharides, those containing a $\beta(1\rightarrow6)$ linkage were the best for supporting growth of bifidobacteria in mixed culture populations (Sanz et al., 2005). In this respect, it was observed that GOS mixtures produced with β-galactosidases from *Lactobacillus reuteri* and *L. plantarum* were rich in these $\beta(1\rightarrow6)$ linked oligosaccharides and should be of interest as prebiotic mixtures (Splechtna et al., 2007; Iqbal et al., 2010).

Although several studies have assessed the ability of different microorganisms to synthesize GOS, only few of them have assayed *in vivo* in animal models or humans the prebiotic effect of the GOS obtained. In a sequence of researches, a novel GOS mixture (B-GOS) designed to specifically improve *Bifidobacterium* numbers was synthesized through the β-galactosidase activity of probiotic *Bifidobacterium bifidum* NCIMB 41171 (Tzortzis et al., 2005a). With the use of an *in vitro* model of the human colon and an *in vivo* pig-feeding trial (Tzortzis et al., 2005b), the increased bifidogenic activity of this novel galactooligosaccharide mixture was demonstrated. In the following study, Depient et al. (2008) assessed in a double-blind, randomized, placebo-controlled, crossover trial, the prebiotic effect of the B-GOS produced by the probiotic *B. bifidum* strain compared with a GOS mixture produced through enzymes from *Bacillus circulans* ATCC 4516. Results showed that the novel B-GOS, as part of Western diet, exerted a bifidogenic effect in a dose-response relation in healthy human volunteers at doses of 3.6 and 7 g/d. Moreover, GOS synthesized by *B. bifidum* showed significantly higher prebiotic effects than did GOS produced with enzymes from *B. circulans*. This study confirmed that GOS produced with different β-galactosidases display different prebiotic properties and propose the usage of enzymes coming from bifidobacterial species in order to increase the bifidogenic properties of prebiotic products.

In conclusion evidences suggest that the development of prebiotic oligosaccharides with enhanced selectivity toward specific bacterial groups is feasible.

Concluding Remarks: Future Researches

Lactose, the sugar of milk, is besides an important energetic fuel for the body of tolerant people a substrate for microbial enzymatic transformation. This bioconversion is carried out by the enzyme β-galactosidase which has been described and characterized in many microorganisms. β-galactosidase is by far one of the most studied enzymes and its activity possess a great relevance in the fields of food and pharmaceutical industries. In this regard, the majority of researches have focused in three main biotechnological applications: *i)* selection of probiotic microorganisms with β-galactosidase active in the gastrointestinal tract to improve lactose maldigestion, *ii)* the usage of the enzyme in whole cells or extracted and purified, (free or immobilized), to hydrolyze lactose or *iii)* to synthesize galactooligo-saccharides (GOS) with prebiotic properties.

Promising results, that encourage further investigations, have been obtained in the three topics. In the case of lactose intolerance, the selection of probiotic microorganisms with appropriate β-galactosidase activities should be complemented with *in vivo* assays of survival and activity in the gut (both in animal models and clinical trials). Besides, certain technological parameters should be assessed in order to include them in the formulation of novel functional foods.

In the same way, numerous evidences support the potential of different β-galactosidases to be used in the hydrolysis or transgalactosilation of lactose in milk or whey. However, much of these results remain at the experimental level so that special attention should be taken in the research of new production processes and in the scaling to an industrial level.

GOS production represents an important but still partially explored field for the development of new prebiotic ingredients for functional foods. These compounds have an important advantage over other prebiotic oligosaccharides as they can be obtained from a low-cost substrate like cheese whey and serve to create high-value products.

It has been observed that depending on the producer microorganism, a wide range of GOS can be obtained. However, studies that clearly relate structure with function (prebiotic effect) are still missing.

Although many researches have been published in the last years, it remains a challenge to sufficiently understand the structure and activity of β-galactosidase, in order to increase the efficiency of hydrolysis or transgalactosylation and to improve the GOS yield and quality of products in a rational way.

Nowadays tendency goes towards the use of genetic engineering for the development of recombinant live β-gal delivery systems using food-grade protein expression techniques and selected probiotics as vehicle; or for the increase of β-gal activity and the production of GOS with a specific degree of polymerization and geometric configuration. Further enzyme engineering could be directed towards enzymes that produce GOS with higher prebiotic efficacy.

However, the use of recombinant technology for obtaining enzymes with enhanced properties or GOS production is still limited. Besides, the negative public opinion towards the inclusion of genetically modified ingredients in food needs to be overcome in order to render these approaches suitable for food and dairy industry.

In the near future, great advances are expected in the development of functional foods containing probiotic microorganisms with high β-galactosidase activity that produces prebiotic GOS with increased selectivity, in order to supply the growing demand of healthy products by consumers.

References

Albayrak N. & Yang S.-T. (2002). Production of galacto-oligosaccharides from lactose by *Aspergillus oryzae* β-galactosidase immobilized on cotton cloth. *Biotechnology and Bioengineering, 77*, 7-19.

Akiyama, K., Takase, M., Horikoshi, K. & Okonogi, S. (2001). Production of galactooligosaccharides from lactose using a beta-glucosidase from *Thermus sp* Z-1. *Bioscience, Biotechnology and Biochemestry*, *65*, 438–441.

Alliende, F. G. (2007). Lactose and other disaccharides intolerance. *Gastroenterología Latinoamericana*, *18*, 152-156.

Arola, H. (1994). Diagnosis of hypolactasia and lactose malabsorption. *Scandinavian Journal of Gastroenterology, 202*, 26-35.

Arrigoni, E., Marteau, P., Briet, F., Pochart, P., Rambaud, J-C. & Messing, B. (1994). Tolerance and absorption of lactose from milk and yogurt during short-bowel syndrome in humans. *American Journal of Clinical Nutrition*, *60*, 926–929.

Audic, J. L., Chaufer, B. & Daufinm, G. (2003). Non-food applications of milk components and dairy coproducts: a review. *Lait*, *83*, 417–438.

Barreteau, H., Delattre, C. & Michaud, P. (2006). Production of oligosaccharides as promising new food additive generation. *Food Technology and Biotechnology*, *44*, 323-333.

Byfield, J., Cárdenas, S., Alméciga-Diaz, C. J. & Sánchez, O. F. (2010). β-Galactosidase and Galactooligosaccharides Production and Applications *Recent Patents on Chemical Engineering, 3*, 17-29.

Boll, W., Wagner, P. & Mantei, N. (1991). Structure of the chromosomal gene and cDNA coding for Lactase phlorizin hydrolase in humans with adult-type hypolactasya or persistance of lactase. *American Journal of Human Genetics*, *48*, 889-902.

Boudraa, G., Touhami, M. R., Pochart, P., Soltana, R., Mary, J. Y. & Desjeux, J. F. (1990). Effect of feeding yogurt versus milk in children with persist diarrhea. *Journal of Pediatric Gastroenterology and Nutrition*, *11*, 509-512.

Briet, F., Pochart, P., Marteau, P., Flourié, B., Arrigoni, E. & Rambaud, J-C. (1997). Improved clinical tolerance to chronic lactose ingestion in subjects with lactose intolerance:a placebo effect?. *Gut*, *41*, 632–635.

Bruins, M. E., Strubel, M., van Lieshout, J. F. T., Janssen, A. E. M. & Boom R. M. (2003). Oligosaccharide synthesis by the hyperthermostable beta-glucosidase from *Pyrococcus furiosus*: kinetics and modelling. *Enzyme and Microbial Technology*, *33*, 3–11.

Büller, H. A., Montgomery, R. K., Sasak, W. V. & Grand, R. J. (1987). Biosynthesis, glycosylation, and intracellular transport of intestinal lactase-phlorizin hydrolase in rat. *Journal of Biological Chemistry*, *262*, 17206-17211.

Cesca, B., Manca De Nadra, M. C., Strasser De Saad, A. M., De Ruiz Holgado, A. & Oliver, G. (1984). β- D-galactosidase of *Lactobacillus* species. *Folia Microbiologica*, *29*, 288-294.

Chassy, B. M., Gibson, E. M. & Giuffrida, A. (1978). Evidence for plasmid-associated lactose metabolism in *Lactococcus* subsp. *casei*. *Current Microbiology*, *1*, 141-144.

Chassy, B. M. & Alpert, C. A. (1989). Molecular organization of the plasmic encoded lactose-PTS of *Lactobacillus casei*. *FEMS Microbiology Reviews*, *63*, 157-166.

Chen, W., Chen, H., Xia, Y., Zhao, J., Tian, F. & Zhang, H. (2008). Production, purification, and characterization of a potential thermostable galactosidase for milk lactose hydrolysis from *Bacillus stearothermophilus*. *Journal of Dairy Science*, *91*, 1751-1758.

Cho, Y. J., Shin, H. J. & Bucke, C. (2003). Purification and biochemical properties of a galactooligosaccharide producing β-galactosidase from *Bullera singularis*. *Biotechnology Letters*, *25*, 2107–2111.

Chockchaisawasdee, S., Athanasopoulos, V. I., Niranjan, K. & Rastall, R. A. (2005). Synthesis of galacto-oligosaccharide from lactose using beta-galactosidase from *Kluyveromyces lactis*: Studies on batch and continuous UF membrane-fitted bioreactors. *Biotechnology and Bioengineering*, *89*(4), 434–443.

Cousin, F. J., Mater, D. D., Foligné, B. & Jan, G. (2010). Dairy propionibacteria as human probiotics: a review of recent evidence. *Dairy Science and Technology*. doi: 10.1051/dst/2010032.

Cruz, R., D'Arcadia, C. V., Belote, J. G., de Oliveira, K. M., Dorta, C., dos Santos O. L. H., Ardiles, E. & Galli A. (1999). Production of transgalactosylated oligosaccharides (TOS) by galactosyltransferase activity from *Penicillium simplicissimum*. *Bioresources Technology, 70*, 165–71.

Dahlqvist, A. (1961). The location of carbohydrases in the digestive tract of the pig. *Biochemical Journal, 78*, 282-288.

Danielsen, E. M., Skovbjerg, H., Noren, O. & Sjostrom H. (1984). Biosynthesis of intestinal microvillar proteins: intracellular processing of lactase-phlorizin hydrolase. *Biochemical and Biophysical Research Communications, 122*, 82-90.

De Vrese, M., Stegelmann, A., Richter, B., Fenselau, S., Laue, C. & Schrezenmeir, J. (2001). Probiotics-compensation for lactase insufficiency. *The American Journal of Clinical Nutrition, 73*, 421S-429S.

Dehkordi, N., Rao, D. R., Warren, A. P. & Chawan, C. B. (1995). Lactose malabsorption as influenced by chocolate milk, skim milk, sucrose, whole milk, and lactic cultures. *Journal of the American Dietetic Association, 95*, 484–486.

Depeint, F., Tzortzis, G., Vulevic, J., I'Anson, K. & Gibson, G. R. (2008). Prebiotic evaluation of a novel galactooligosaccharide mixture produced by the enzymatic activity of *Bifidobacterium bifidum* NCIMB 41171, in healthy humans: a randomized, double-blind, crossover, placebo-controlled intervention study. *American Journal of Clinical Nutrition, 87*, 785–791.

Di Stefano, M., Veneto, G., Malservisi, S., Cecchetti, L., Minguzzi, L., Strocchi, A. & Corazza, G. R. (2002). Lactose malabsorption and intolerance and peak bone mass. *Gastroenterology, 122*, 1793-1799.

Doell, R. G. & Kretchmer, N. (1962). Studies of small intestine development.1. Distribution and activity of P-galactosidase. *Biochimica et Biophysica Acta, 62*, 353-362.

Dumortier, V., Montreuil, M. & Bouquelet, S. (1990). Primary structure of ten galactosides formed by transglycosylation during lactose hydrolysis by Bifidobacterium bifidum. *Carbohydrate Research, 201*, 115-123.

Falentin, H., Deutsch, S. M., Jan, G., Loux, V., Thierry, A. & Parayre, S. (2010). The complete genome of *Propionibacterium freudenreichii* CIRM-BIA1, a hardy actinobacterium with food and probiotic applications. *PLoS One. 5*(7), e11748.

FAO/WHO. (2002). Probiotics in Foods. Health and nutritional properties and guidelines for evaluation. Food and Health Agricultural Organization of the United Nations and World Health Organization Working Group Report. ftp://ftp.fao.org/docrep/fao/009/a0512e/a0512e00.pdf

Fernandez, M., Margolles, A., Suarez, J. E. & Mayo, B. (1999). Duplication of the β-galactosidase gene in some *Lactobacillus plantarum* strains. *International Journal of Food Microbiology*, *48*, 113-123.

Flatz, G. (1995). The genetic polymorphism of intestinal lactase activity in adult humans. In: Scriver C.R., Beaudet A.L. & Sly W.S. (Eds.) The metabolic and molecular basis of Inherited Disease, (7th ed., 4441-4450). London, UK: *McGraw-Hill.*

Fox, P. F., Lucey, J.A. & Cogan, T. M. (1990). Glycolysis and related reactions during cheese manufacture and ripening. *Critical Reviews in Food Science and Nutrition*, *29*, 237-253.

Fuller, R. & Gibson, G. R. (1998). Probiotics and prebiotics: microflora management for improved gut health. *Clincinal Microbiology and Infection*, *4*, 477–480.

Gallagher, C. R., Molleson, A. L. & Caldwell, J. H. (1974). Lactose intolerance and fermented dairy products. *Journal of the American Dietetics Association*. *65*, 418-419.

Gänzle, M. (2012). Enzymatic synthesis of galacto-oligosaccharides and other lactose derivatives (hetero-oligosaccharides) from lactose. *International Dairy Journal*, *22*, 116-122.

García-Garibay, M. & Gómez-Ruiz, L. (1996). Usos de β-galactosidasas microbianas para reducir el contenido de lactosa en leche y productos lácteos. *Revista de Investigación Clínica*, *48*, 51-56.

Gasson, M. J., Hill, S. H. A. & Anderson, P. H. (1987). Molecular genetics of metabolic traits in lactic streptococci. In Ferretti J.J. & Curtiss, R (Eds.) Streptococcal Genetics. (1st ed., 242-245.) Washington, D.C., USA: *American Society of Microbiology*.

Gaur, R., Pant, H., Jain, R. & Khare, S. K. (2006). Galacto-oligosaccharide synthesis by immobilized *Aspergillus oryzae* beta-galactosidase. *Food Chemistry*, *97*, 426–430.

Gilat, T., Russo, S., Gelman-Malachi, E. & Aldor, T. A. (1972). Lactase in man: a nonadaptable enzyme. *Gastroenterology*, *62*, 1125-1127.

Gilliland, S. E. & Kim, S. H. (1984). Effect of viable starter culture bacteria in yogurt on lactose utilization in humans. *Journal of Dairy Science*, *67*, 1-6.

Gosling, A., Alftren, J., Stevens, G. W., Barber, A. R., Kentish, S. E. & Gras, S. L. (2009). Facile pretreatment of *Bacillus circulans* beta galactosidase increases the yield of galactosyl oligosaccharides in milk and lactose reaction systems. *Journal of Agricultural and Food Chemistry*, *57*, 11570-11574.

Gosling, A., Stevens, G. W., Barber, A. R., Kentish, S. E. & Gras, S. L. (2010). Recent advances refining galactooligosaccharide production from lactose. *Food Chemistry*, *121*, 307–318.

Goulas, T. K., Goulas, A. K., Tzortzis, G. & Gibson, G. R. (2007). Molecular cloning and comparative analysis of four beta-galactosidase genes from *Bifidobacterium bifidum* NCIMB41171. *Applied Microbiology and Biotechnology*, *76*, 1365–1372.

Grosová, Z., Rosenberg, M. & Rebroš, M. (2008). Perspectives and Applications of Immobilised β-Galactosidase in Food Industry – a Review. *Czech Journal of Food Science*, *26*, 1–14.

Haider, T. & Husain, Q. (2009). Immobilization of beta galactosidase *from Aspergillus oryzae* via immunoaffinity support. *Biochemical Engineering Journal*, *43*, 307–314.

Hansson, T., Andersson, M., Wehtj, E. & Adlercreutz, P. (2001a). Influence of water activity on the competition between b-glycosidase-catalysed transglycosylation and hydrolysis in aqueous hexanol. *Enzyme and Microbial Technology*, *29*, 527–534.

Hansson, T., Kaper, T., van der Oost, J., de Vos, W. M. & Adlercreutz, P. (2001b). Improved oligosaccharide synthesis by protein engineering of beta-glucosidase CelB from hyperthermophilic *Pyrococcus furiosus*. *Biotechnology and Bioengineering*, *73*, 203 -210.

Hansson, T. & Adlercreutz, P. (2001c). Optimization of galactooligosaccharide production from lactose using beta-glycosidases from hyperthermophiles. *Food Biotechnology*, *15*, 79–97.

Hartley, J. C. & Vedamuthu, E. R. (1975). ß-Galactosidase of *Propionibacterium shermanii. Applied Microbiology and Biotechnology*, *29*, 74-80.

He, M., Antoine, J. M., Yang, Y., Yang, J., Men, J. & Han, H. (2004). Influence of live flora on lactose digestion in male adult lactose-malabsorbers after daily products intake. *Wei Sheng Yan Jiu – Journal of Hygiene Research*, *33*, 603–605.

He, T., Priebe, M. G., Zhong, Y., Huang, C., Harmsen, H. J., Raangs, G. C., Antoine, J. M., Welling, G. W. & Vonk, R. J. (2008). Effects of yogurt and bifidobacteria supplementation on the colonic microbiota in lactose-intolerant subjects. *Journal of Applied Microbiology*, *104*, 595–604.

Hemme, D., Nardi, M. & Jctte J. (1980). Beta-galactosidasa et phospho-beta-galactosidases de *Streptococcus thermophilus*. *Le lait*, LX, 595-618.

Hertzler, S. R. & Savaiano, D. A. (1996). Colonic adaptation to daily lactose feeding in lactose maldigesters reduces lactose intolerance. *American Journal of Clinical Nutrition*, *64*, 232–236.

Hickey, M. W., Hillier, A. J. & Jago, G. R. (1986). Transport and metabolism of lactose, glucose and galactose in homofermentative lactobacilli. *Applied and Environmental Microbiology*, *51*, 825-831.

Hinz, S. W. A., van den Broek, L. A. M., Beldman, G., Vincken, J.-P. & Voragen, A. G. J. (2004). β-Galactosidase from *Bifidobacterium adolescentis* DSM20083 prefers β(1,4)-galactosides over lactose. *Applied Microbiology and Biotechnology*, *66*, 276-284.

Hove, H., Nordgaard-Andersen, I. & Mortensen, P. B. (1994). Effect of lactic acid bacteria on the intestinal production of lactate and short-chain fatty acids and the absorption of lactose. *American Journal of Clinical Nutrition*, *59*, 74–79.

Hsu, C. A., Yu, R. C. & Chou, C. C. (2005). Production of β-galactosidase by Bifidobacteria as influenced by various culture conditions. *International Journal of Food Microbiology*, *104*, 197–206.

Hsu, C. A., Lee, S. L & Chou, C. C. (2007). Enzymatic Production of Galactooligosaccharides by *β*-Galactosidase from *Bifidobacterium longum* BCRC 15708. *Journal of Agriculture and Food Chemistry*, *55*, 2225-2230.

Huber, R. E., Kurz, G. & Wallenfels, K. (1976). A quantitation of the factors which affect the hydrolase and transgalactosylase activities of β-galactosidase (*E. coli*) on lactose. *Biochemistry, 15*, 1994-2001.

Huang, D. Q., Prevost, H. & Divies, C. (1995). Principal characteristics of β-galactosidase from *Leuconostoc* spp. *International Dairy Journal*, *5*, 29-43.

Hung, M.-N., Xia, Z., Hu, N. T. & Lee, B. (2001). Molecular and biochemical analysis of two β-galactosidases from *Bifidobacterium infantis* HL96. *Applied and Environmental Microbiology*, *67*, 4256-4263.

Hutkins, R. W. & Poone, C. (1991). Lactose uptake driven by galactose efflux in *Streptococcus thermophilus*: evidence for galactose-lactose antiporter. *Applied and Environmental Microbiology*, *57*, 941-944.

Iqbal, S., Nguyen, T. H., Nguyen, T. T. Maischberger, T. & Haltrich, D. (2010). β-Galactosidase from *Lactobacillus plantarum* WCFS1: biochemical characterization and formation of prebiotic galacto-oligosaccharides. *Carbohydrate Research*, *345*, 1408–1416.

Ito, M. & Kimura, M. (1993). Influence of lactose on faecal microflora in lactose maldigestors. *Microbial Ecology in Health and Disease*, *6*, 73–76.

Itoh, T., Ohhashi, M., Toba, T. & Adachi, S. (1980). Purification and properties of β-galactosidase from *Lactobacillus bulgaricus*. *Milchwissenschaft*, *35*, 593-597.

Iwasaki, K., Nakajima, M. & Nakao, S. (1996). Galacto-oligosaccharide production from lactose by an enzymatic batch reaction using b-galactosidase. *Process Biochemistry*, *31*, 69–76.

Ji, E. S., Park, N. H. & Oh, D. K. (2005). Galacto-oligosaccharide production by a thermostable recombinant beta-galactosidase from *Thermotoga maritima*. *World Journal of Microbiology and Biotechnology*, *21*, 759–764.

Jiang, T., Mustapha, A. & Savaiano, D. (1996). Improvement of lactose digestion in humans by ingestion of unfermented milk containing *Bifidobacterium longum*. *Journal of Dairy Science*, *79*, 750-757.

Jiang, T. & Savaiano, D. A. (1997a). In vitro lactose fermentation by human colonic bacteria is modified by *Lactobacillus acidophilus* supplementation. *Journal of Nutrition*, *127*, 1489–1495.

Jiang, T. & Savaiano, D. A. (1997b). Modification of colonic fermentation by bifidobacteria and pH in vitro. Impact on lactose metabolism, short-chain fatty acid, and lactate production. *Digestive Diseases and Sciences*, *42*, 2370–2377.

Jorgensen, F., Hansen, O. C. & Stougaard, P. (2001). High-efficiency synthesis of oligosaccharides with a truncated β-galactosidase from *Bifidobacterium bifidum*. *Applied Microbiology and Biotechnology*, *57*, 647-652.

Jørgensen, F., Hansen, O. C. & Stougaard, P. (2007). Beta-galactosidase isolated from bifidobacterium. Patent no. EP1283876.

Kanatani, K., Tahara, T., Yoshida, K., Miura, H., Sakamoto, M. & Oshimura, M. (1992). Plasmid-linked galactose utilization by *Lactobacillus acidophilus* TK8912, *Bioscience, Biotechnology and Biochemistry*, *56*, 826-827.

Kandler, O. (1983). Carbohydrate metabolism in lactic acid bacteria. *Antoine van Leeuwenhoek*, *49*, 209-224.

Klein, M. P., Fallavena, L. P., Schöffer, J. N., Ayubb, M. A. Z., Rodrigues, R. C., Ninowa, J. L. & Hertz, P. F. (2013). High stability of immobilized β-d-galactosidase for lactose hydrolysis and galactooligosaccharides synthesis. *Carbohydrate Polymers*, *95*, 465– 470.

Kim, H. S. & Gilliland, S. E. (1983). *Lactobacillus acidophilus* as a dietary adjunct for milk to aid lactose digestion in humans. *Journal of Dairy Science*, *66*, 959 – 966.

Kolars, J. C., Levitt, M. D., Aouji, M. & Savaiano, D. A. (1984). Yogurt – an autodigesting source of lactose. *The New England Journal of Medicine*, *310*, 1–3.

Kotz, C. M., Furne, J. K., Savaiano, D. A. & Levitt, M. D. (1994). Factors affecting the ability of a high beta-galactosidase yogurt to enhance lactose absorption. *Journal of Dairy Science*, *77*, 3538–3544.

Krzewinski, F., Brassart, C., Gavini, F. & Bouquelet, S. (1996). Characterization of the Lactose transport system in the strain *Bifidobacterium bifidum* DSM 20082. *Current Microbiology*, *32*, 301-307.

Langsrud, T. & Reinbold, G. W. (1973). Flavour development and microbiology of Swiss cheese - a review. III. Ripening and flavour production. *Journal of Milk and Food Technology*, *36*, 593-609.

Lee, M. F. & Krasinski, S. D. (1998). Human adult-onset lactase decline: an update. *Nutrition Reviews*, *56*, 1-8.

Levri, K. M., Ketvertis, K., Deramo, M., Merenstein, J. H. & D'Amico, F. (2005). Do probiotics reduce adult lactose intolerance? A systematic review. *The Journal of Family Practice*, *54*, 613– 620.

Li, J., Zhang, W., Wang, C., Yu, Q., Dai, R. & Pei, X. (2012). *Lactococcus lactis* expressing food-grade β-galactosidase alleviates lactose intolerance symptoms in post-weaning Balb/c mice. *Applied Microbiology and Biotechnology*, *96*, 1499-1506.

Lin, W. J., Savaiano, D. & Harlander, S. (1991). Influence of non-fermented dairy products containing bacterial starter cultures on lactose maldigestion in humans. *Journal of Dairy Science*, *74*, 87-95.

Macfarlane, G. T., Steed, H. & Macfarlane, S. (2008). Bacterial metabolism and health-related effects of galacto-oligosaccharides and other prebiotics. *Journal of Applied Microbiology*, *104*, 305-344.

Mackie, R. I., Sghir, A. & Gaskins, H. R. (1999). Developmental microbial ecology of the neonatal gastrointestinal tract. *American Journal of Clinical Nutrition*, *69*, 1035S–1045S.

Mahoney, R. (1998). Galactosyl-oligosaccharide formation during lactose hydrolysis: a review. *Food Chemistry*, *63*, 147 – 154.

Mantei, N., Villa, M., Enziler, T., Wacker, H., Boll, W., James, P., Hunziker, W. & Semenza, G. (1988). Complete primary structure of human and rabbit lactase-phlorizin hydrolase: implications for biosynthesis, membrane anchoring and evolution of the enzyme. *EMBO Journal, 7*, 2705-2713.

Marteau, P., Vesa, T. & Rambaud, J. C. (1997). Lactose maldigestion. In: Fuller R. (Eds.) Probiotics 2: applications and practical aspects (65 – 88). London, UK: *Chapman & Hall.*

Martinez-Villaluenga, C., Cardelle-Cobas, A., Corzo, N., Olano, A., & Villamiel, M. (2008). Optimization of conditions for galactooligosaccharide synthesis during lactose hydrolysis by β-galactosidase from *Kluyveromyces lactis* (Lactozym 3000 L HP G). *Food Chemistry, 107*, 258–264

Martini, M. C., Bollweg, G. L., Levitt, M. D., Savaiano, D. A. (1987a). Lactose digestion by yogurt beta-galactosidase: influence of pH and microbial cell integrity. *American Journal of Clinical Nutrition*, *45*, 432–436.

Martini, M. C., Smith, D. E., Savaiano, D. A. (1987b). Lactose digestion from flavored and frozen yogurts, ice milk, and ice cream by lactase-deficient persons. *American Journal of Clinical Nutrition*, *46*, 636–640.

Martini, M. C. & Savaiano, D. A. (1988). Reduced intolerance symptoms from lactose consumed during a meal. *American Journal of Clinical Nutrition*, *47*, 57-60.

Martini, M. C., Lerebours, E. C., Lin, W. J., Harlander, S. K., Berrada, N. M., Antoine, J. M. & Savaiano, D. A. (1991). Strains and species of lactic acid bacteria in fermented milks (yogurts): effect on in vivo lactose digestion. *American Journal of Clinical Nutrition*, *54*, 1041–1046.

Massi, M., Vitali, B., Federici, F., Matteuzzi, D. & Brigidi, P. (2004). Identification method based on PCR combined with automated ribotyping for tracking probiotic Lactobacillus strains colonizing the human gut and vagina. *Journal of Applied Microbiology*, *96*, 777–786.

Mateo, C., Palomo, J. M., Fernandez-Lorente, G., Guisan, J. M & Fernandez-Lafuente, R. (2007). Improvement of enzyme activity, stability and selectivity via immobilization techniques. *Enzyme and Microbial Technology*, *40*, 1451– 1463.

McDonough, F. E., Hitchins, A. D., Wong, N. P., Wells, P. & Bodwell, C. E. (1987). Modification of sweet acidophilus milk to improve utilization by lactose-intolerant persons. *American Journal of Clinical Nutrition*, *45*, 570-574.

Mlichova, Z. & Rosenberg, M. (2006). Current trends of b-galactosidase application in food technology. *Journal of Food and Nutrition Research*, *45*, 47–54.

Møller, P. L., Jørgensen, F., Hansen, O. C., Madden, S. M. & Stougaard, P. (2001). Intra- and Extracellular b-Galactosidases from *Bifidobacterium bifidum* and *B. infantis*: Molecular Cloning, Heterologous Expression, and Comparative Characterization *Applied and Environmental Microbiology*, *67*, 2276–2283.

Montalto, M., Curigliano, V., Santoro, L., Vastola, M., Cammarota, G., Manna, R., Gasbarrini, A. & Gasbarrini, G. (2006). Management and treatment of lactose malabsorption. *World Journal of Gastroenterology*, *14*, 187-191.

Montes, R. G., Bayles, T. M., Saavedra, J. M. & Perman, J. A. (1995). Effect of milks Inoculated with *Lactobacillus acidophilus* or a yogurt starter culture in lactose Maldigesting children. *Journal of Dairy Science*, *78*, 1657-1664.

Morelli, L. & Capurso, L. (2012). FAO/WHO guidelines on probiotics: 10 years later. *Journal of Clinical Gastroenterology*, *46* Suppl., S1-S2.

Mozaffar, Z., Nakanishi, K., Matsuno, R. & Kamikubo, T. (1984). Purification and properties of b-galactosidases from *Bacillus circulans*. *Agricultural and Biological Chemistry*, *48*, 3053–3061.

Mozaffar, Z., Nakanishi, K., Matsuno, R. & Kamikubo, T. (1984). Purification and properties of beta-galactosidases from Bacillus circulans. *Agricultural and Biological Chemistry, 48*(12), 3053–3061.

Mustapha, A., Jiang, T. & Savaiano, D. (1997). Improvement of lactose digestion by humans following ingestion of unfermented acidophilus milk: influence of bile sensitivity, lactose transport, and acid tolerance of *Lactobacillus acidophillus*. *Journal of Dairy Science*, *80*, 1537-1545.

Naim, H. Y., Sterchi, E. E. & Lentze, M. J. (1987). Biosynthesis and maturation of lactase-phlorizin hydrolase in human small intestinal epithelial cells. *Biochemical Journal*, *241*, 427-434.

Nakagawa, T., Fujimoto, Y., Ikehata, R., Miyaji, T. & Tomizuka, N. (2006). Purification and molecular characterization of cold-active β-galactosidase from *Arthrobacter psychrolactophilus* strain F2. *Applied Microbiology and Biotechnology*, *72*, 720-725.

Nakao, M., Harada, M., Kodama, Y., Nakayama, T., Shibano, Y. & Amachi, T. (1994). Purification and characterization of a thermostable beta-galactosidase with high transgalactosylation activity from *Saccharopolyspora rectivirgula*. *Applied Microbiology and Biotechnology*, *40*, 657–663.

Neri, D. F. M., Balcao, V. M., Costa, R. S., Rocha, I. C., Ferreira, E. M. F. C. & Torres, D. P. M., et al. (2009). Galacto-oligosaccharides production during lactose hydrolysis by free Aspergillus oryzae b-galactosidase and immobilized on magnetic polysiloxane-polyvinyl alcohol. *Food Chemistry*, *115*, 92-99.

Newcomer, A. D., Park, H. S., O'Brien, P. C. & McGill, D. B. (1983). Response of patients with irritable bowel syndrome and lactase deficiency using unfermented acidophilus milk. *American Journal of Clinical Nutrition*, *38*, 257–263.

Nguyen, T. H., Splechtna, B., Steinböck, M., Kneifel, W., Lettner, H. P., Kulbe, K. D. & Haltrich, D. (2006). Purification and characterization of two novel betagalactosidases from *Lactobacillus reuteri*. *Journal of Agricultural and Food Chemistry*, *54*, 4989–4998.

Nilsson, T. K. & Johansson, C. A. (2004). A novel method for diagnosis of adult hypolactasia by genotyping of the -13910 C/T polymorphism with Pyrosequencing technology. *Scandinavian Journal of Gastroenterology*, *39*, 287-290.

Noh, D. O. & Gilliand, S. E. (1993). Influence of bile on cellular integrity and beta-galactosidase activity of Lactobacillus acidophilus. *Journal of Dairy Science*, *76*, 1253-1259.

Noh, D. O. & Gilliand, S. E. (1994). Influence of bile on beta-galactosidase activity of component species of yogurt starter cultures. *Journal of Dairy Science*, *77*, 3532–3537.

Noren, O. & Sjøstriim, H. G. (2001). Structure, biosynthesis and regulation of lactase-phlorizin hydrolase. *Scandinavian Journal of Nutrition*, *45*, 156-160.

O'Connell, S. & Walsh, G. (2010). A novel acid-stable, acid-active β-galactosidase potentially suited to the alleviation of lactose intolerance. *Applied Microbiology and Biotechnology*, *86*, 517–524.

Onishi, N. & Tanaka, T. (1995). Purification and properties of a novel thermostable galacto-oligosaccharide-producing beta-galactosidase from *Sterigmatomyces elviae* CBS8119. *Applied and Environmental Microbiology*, *61*, 4026–4030.

Onishi, N., Yamashiro, A. & Yokozeki, K. (1995). Production of galacto-oligosaccharide from lactose by *Sterigmatomyces elviae* CBS8119. *Applied and Environmental Microbiology*, *61*, 4022–4025.

Onishi, N. & Yokozeki, K. (1996). Gluco-oligosaccharide and galacto-oligosaccharide production by *Rhodotorula minuta* IFO879. *Journal of Fermentation and Bioengineering*, *82*, 124–127.

Onishi, N. & Tanaka, T. (1997). Purification and characterization of galacto-oligosaccharide-producing beta-galactosidase from *Sirobasidium magnum*. *Letters in Applied Microbiology*, *24*, 82–86.

Onwulata, C. I., Rao, D. R. & Vankineni, P. (1989). Relative efficiency of yogurt, sweet acidophilus milk, hydrolyzed-lactose milk, and a commercial lactase tablet in alleviating lactose maldigestion. *American Journal of Clinical Nutrition*, *49*, 1233-1237.

Otieno, D. O. (2010). Synthesis of b-galactooligosaccharides from lactose using microbial b-galactosidases. *Comprehensive Reviews in Food Science and Food Safety*, *9*, 471–482.

Osman, A., Tzortzis, G., Rastall, R. A. & Charalampopoulos, D. (2010). A comprehensive investigation of the synthesis of prebiotic galactooligosaccharides by whole cells of *Bifidobacterium bifidum* NCIMB 41171. *Journal of Biotechnology*, *150*, 140–148.

Panesar, P. S., Kumari, S. & Panesar, R. (2010). Potential Applications of Immobilized β-Galactosidase in Food Processing Industries. *Enzyme Research*, Article ID 473137, 16, doi:10.4061/2010/473137.

Park, H. Y., Kim, H. J., Lee, J. K., Kim, D. & Oh, D. K. (2008). Galactooligosaccharide production by a thermostable beta-galactosidase from Sulfolobus solfataricus. *World Journal of Microbiology and Biotechnology*, *24*, 1553–1558.

Pessela, B. C. C., Mateo, C., Carrascosa, A. C., Vian, A., García, J. L., Rivas, G., Alfonso, C., Guisan, J. M. & Fernández-Lafuente, R. (2003). One-step purification, covalent immobilization, and additional stabilization of a thermophilic poly-his-tagged β-galactosidase from *Thermus* sp. strain T2 by using novel heterofunctional chelate-epoxy sepabeads. *Biomacromolecules*, *4*, 107–113.

Perez Chaia, A., Zarate, G. & Oliver, G. (1999). The probiotic properties of *propionibacteria*. *Lait*, *79*, 175-185.

Perez Chaia, A. & Zárate, G. (2005). Dairy propionibacteria from milk or cheese diets remain viable and enhance propionic acid production in the mouse caecum. *Lait*, *85*, 85-98.

Placier, G., Watzlawick, H., Rabiller, C. & Mattes, R. (2009). Evolved betagalactosidases from *Geobacillus stearothermophilus* with improved transgalactosylation yield for galacto-oligosaccharide production. *Applied and Environmental Microbiology*, *75*, 6312–6321.

Petzelbauer, I., Kuhn, B., Splechtna, B., Kulbe, K. D. & Nidetzky, B. (2002). Development of an ultrahigh-temperature process for the enzymatic hydrolysis of lactose. IV. Immobilization of two thermostable bglycosidases and optimization of a packed-bed reactor for lactose conversion. *Biotechnology and Bioengineering*, *77*, 619-631.

Prenosil, J. E., Stuker, E. & Bourne, J. R. (1987). Formation of oligosaccharides during enzymatic lactose hydrolysis and their importance in a whey hydrolysis process: Part II: Experimental. *Biotechnology and Bioengineering*, *30*, 1026–1031.

Quesada-Chanto, A., Afschar, A. S. & Wagner, F. (1994). Microbial production of propionic acid and vitamin B12 using molasses or sugar. *Applied Microbiology and Biotechnology*. 41, 378-383.

Rabiu, B. A., Jay, A. J., Gibson, G. R. & Rastall, R. A. (2001). Synthesis and fermentation properties of novel galacto-oligosaccharides by beta-galactosidases from *Bifidobacterium* species. *Applied and Environmental Microbiology*, *67*, 2526– 2530.

Ramana, M. V. & Dutta, S. M. (1978). Lactase Activity of Microorganisms. *Folia Microbiologica*, *23*, 210-215.

Rambaud, J. C., Bouhnick, Y. & Marteau, P. (1994). Dairy products and intestinal flora. In: Serrano-Rios M., Sastre A., Prerez Jeuz M.A., Estrala A. & de Sebastian C. (Eds.) Dairy Products in Human Health and Nutrition (1st Ed., 389-399.) Rotterdam, NL: *A.A. Blakema.*

Rand, A. G. (1981). Enzyme technology and the development of lactose hydrolyzed milk. In: Paige D.M. & Bayless T.M. (Eds.) Lactose Digestion: Clinical and Nutritional Implications. (Illustrated Ed., 219-230) Baltimore, USA: *Johns Hopkins University Press*.

Rasinpera, H., Savilahti, E., Enattah, N. S., Kuokkanen, M., Totterman, N., Lindahl, H., Jarvela, I. & Kolho, K. L. (2004). A genetic test which can be used to diagnose adult-type hypolactasia in children. *Gut* , *53*, 1571-1576.

Rastall, R. A. & Maitin, V. (2002). Prebiotics and synbiotics: towards the next generation. *Current Opinion in Biotechnology*, *13*, 490–496.

Reuter, S., Rusborg, A. & Zimmermann, W. (1999). Galactooligosaccharide synthesis with β-galactosidases from *Sulfolobus solfataricus*, *Aspergillus oryzae* and *Escherichia coli*. *Enzyme and Microbial Technology*, *25*, 509-516.

Rhimi, M., Aghajari, N., Jaouadi, B., Juy, M, Boudebbouze, S., Maguin, E., Haser, R. & Bejar, S. (2009). Exploring the acidotolerance of b-galactosidase from *Lactobacillus delbrueckii* subsp. *bulgaricus*: an attractive enzyme for lactose bioconversion. *Research in Microbiology*, *160*, 775-784.

Roberfroid, M. (2007). Prebiotics: The Concept Revisited. *The Journal of Nutrition*, *137*, 830-837.

Rosado, J. L., Solomons, N. W., Lisker, R. & Bourges, H. (1984). Enzyme replacement therapy for primary adult lactase deficiency. Effective reduction of lactose malabsorption and milk intolerance by direct addition of beta-galactosidase to milk at mealtime. *Gastroenterology*, *87*, 1072-1082.

Rossi, M., Maiuri, L., Fusco, M. I., Salvati, V. M., Fuccio, A., Auricchio, S., Mantei, N., Zecca, L., Gloor, S. M. & Semenza, G. (1997). Lactase persistence versus decline in human adults: multifactorial events are involved in down-regulation after weaning. *Gastroenterology*, *112*, 1506-1514.

Sahi, T. (1994). Genetics and epidemiology of adult-type hypolactasia. *Scandinavian Journal of Gastroenterology*, *29*, Sup. 202, 7-20.

Sako, T., Matsumoto, K. & Tanaka, R. (1999). Recent progress on research and applications. of non-digestible galacto-oligosaccharides. *International Dairy Journal*, *9*, 69 – 80.

Saltzman, J. R., Russell, R. M., Golner, B., Barakat, S., Dallal, G. E. & Goldin, B. R. (1999). A randomized trial of *Lactobacillus acidophilus* BG2FO4 to treat lactose intolerance. *American Journal of Clininical Nutrition*, *69*,140–146.

Sanz, M. L., Gibson, G. R. & Rastall, R. A. (2005). Influence of disaccharide structure on prebiotic selectivity in vitro. *Journal of Agricultural and Food Chemistry*, *53*, 5192–5199.

Savaiano, D. A., AbouElAnouar, A., Smith, D. E. & Levitt, M. D. (1984). Lactose malabsorption from yogurt, pasteurized yogurt, sweet acidophilus milk, and cultured milk in lactase-deficient individuals. *American Journal of Clinical Nutrition*, *40*, 1219–1223.

Savaiano, D. A. & Levitt, M. D. (2000). Lactose Intolerance. In: Miller G. D., Jarvis J.K., & McBean L.D. (Eds.) Handbook of Dairy Foods and Nutrition. (2nd Ed., 311-354) Rosemont, Illiniois, USA: *CRC Press LLC*.

Semenza, G., Auricchio, S. & Mantei, N. (1999). Small-Intestinal Disaccharidases. In: Scriver C.R., Beaudet A.L., Sly W.S. & Valle D. (Eds.) The metabolic and molecular basis of inherited diseases. (8th ed., 1623-1650). New York, USA: *McGraw-Hill*.

Shah, N., Fedorak, R. N. & Jelen, P. (1992). Food consistency effects of quarg in lactose malabsorption. *International Dairy Journal*, *2*, 257-269.

Shaw, A. D. & Davies, G. J. (1999). Lactose intolerance: problems in diagnosis and treatment. *Journal of Clinical Gastroenterology*, *28*, 208-216.

Sherman, P. M. (2004). Probiotics and lactose maldigestion. *The Canadian Journal of Gastroenterology*, *18*, 81-82.

Shimizu-Kadota, M. (1987). Properties of lactose plasmid pLY101 in *Lactobacillus casei. Applied and Environmental Microbiology*, *53*, 2987-2991.

Shin, H. J., Park, J. M. & Yang, J. W. (1998). Continuous production of galacto-oligosaccharides from lactose by *Bullera singularis* β-galactosidase immobilized in chitosan beads. *Process Biochemistry*, *33*, 787–792.

Sinclair, H. R., De Slegte, J., Gibson, G. R. & Rastall, R. A. (2009). Galactooligosaccharides (GOS) inhibit vibrio cholera toxin binding to its GM1 receptor. *Journal of Agricultural and Food Chemistry*, *57*, 3113-3119.

Solomons, N. W., Guerrero, A. M. & Torun, B. (1985). Effective in vivo hydrolysis of milk lactose by beta-galactosidases in the presence of solid foods. *American Journal of Clinical Nutrition*, *41*, 222-227.

Splechtna, B., Nguyen, T. H. & Haltrich, D. (2006). Production of Prebiotic Galacto- Oligosaccharides from Lactose using β-Galactosidases from *Lactobacillus reuteri. Journal of Agricultural and Food Chemistry*, *54*, 4999-5006.

Spletchna, B., Nguyen, T. H. & Haltrich, D. (2007). Comparison between discontinuous and continuous lactose conversion processes for the production of prebiotic galacto-oligosaccharides using beta-galactosidase from *Lactobacillus reuteri. Journal of Agricultural and Food Chemistry*, *55*, 6772-6777.

Tannock, G. W., Munro, K., Harmsen, H. J., Welling, G. W., Smart, J. & Gopal, P. K. (2000). Analysis of the fecal microflora of human subjects consuming a probiotic product containing *Lactobacillus rhamnosus* DR20. *Applied and Environmental Microbiology*, *66*, 2578–2588.

Thompson, J. (1979). Lactose metabolism in *Streptococcus lactis*: phosphorilation of galactose and glucose moieties in vivo. *Journal of Bacteriology*, *144*, 672-682.

Toba, T., Tomita, Y., Itoh, T. & Adachi, S. (1981). β-Galactosidases of lactic acid bacteria: characterization by oligosaccharides formed during hydrolysis of lactose. *Journal of Dairy Science*, *64*, 185-192.

Trincone, A. & Giordano, A. *(*2006*)*. Glycosyl hydrolases and glycosyltransferases in the synthesis of oligosaccharides. *Current Organic Chemestry, 10,* 1163–1169.

Tymczyszyn, E. E., Gerbino, E., Illanes, A. & Gómez-Zavaglia, A. (2011). Galacto-oligosaccharides as protective molecules in the preservation of *Lactobacillus delbrueckii* subsp. *bulgaricus*. *Cryobiology*, *62*, 123-129.

Tzortzis, G., Goulas, A. K. & Gibson, G. R. (2005a). Synthesis of prebiotic galactooligosaccharides using whole cells of a novel strain, *Bifidobacterium bifidum* NCIMB 41171. *Applied Microbiology and Biotechnology*, *68*, 412–416.

Tzortzis, G., Goulas, A. K., Gee, J. M. & Gibson, G. R. (2005b). A novel galactooligosaccharide mixture increases the bifidobacterial population numbers in a continuous in vitro fermentation system and in the proximal colonic contents of pigs in vivo. *Journal of Nutrition*, *135*, 1726–1731.

Varela-Moreiras, G., Antoine, J. M., Ruiz-Roso, B. & Varela, G. (1992). Effects of yogurt and fermented-then-pasteurized milk on lactose absorption in an institutionalized elderly group. *Journal of the American College of Nutrition*, *11*, 168–171.

Vesa, T. H., Korpela, R. A. & Sahi, T. (1996). Tolerance to small amounts of lactose in lactose maldigesters. *American Journal of Clinical Nutrition*, *64*, 197–201.

Vonk, R. J., Lin, Y., Koetse, H. A., Huang, C., Zeng, G., Elzinga, H., Antoine, J. & Stellaard, F. (2000). Lactose (mal)digestion evaluated by the 13C-lactose digestion test. *European Journal of Clinical Investigation*, *30*, 140-146.

Vulevic, J., Drakoularakou, A., Yaqoob, P., Tzortzis, G. & Gibson, G. R. (2008). Modulation of the fecal microflora profile and immune function by a novel trans-galactooligosacharide mixture (B-GOS) in healthy elderly volunteers. *American Journal of Clinical Nutrition*, *88*, 1438–1446.

Wang, C., Zhang, C. W., Liu, H. C., Yu, Q. & Pei, X. F. (2008). Non-fusion and fusion expression of β-galactosidase from *Lactobacillus bulgaricus* in *Lactococcus lactis*. *Biomedical and Environmental Sciences*, *21*, 389–397.

Wang, Y., Harvey, C. B., Hollox, E. J., Phillips, A. D., Poulter, M., Clay, P., Walker-Smith, J. A. & Swallow, D. M. (1998) The genetically programmed down-regulation of lactase in children. *Gastroenterology*, *114*, 1230-1236.

Wang, G., Gao, Y., Hu, B., Lu, X., Liu, X. & Jiao, B. (2013). A novel cold-adapted β-galactosidase isolated from Halomonas sp.S62: gene cloning, purification and enzymatic characterization *World J Microbiol Biotechnol* DOI 10.1007/s11274-013-1311-7.

Wu, Y., Yuan, S., Chen, S., Wu, D., Chen, J. & Wu, J. (2013). Enhancing the production of galacto-oligosaccharides by mutagenesis of *Sulfolobus solfataricus* β-galactosidase. *Food chemistry*, *138*, 1588-1595.

Yanahira, S., Yabe, Y., Nakakoshi, M., Miura, S., Matsubara, N. & Ishikawa, H. (1998). Structures of novel acidic galactooligosaccharides sythesised by *Bacillus circulans* β-galactosidase. *Bioscience, Biotechnology and Biochemistry*, *62*, 1791-1794.

Zárate, G., González, S., Pérez Chaia, A. & Oliver, G. (2000a). Effect of bile on the b-galactosidase activity of dairy propionibacteria. *Lait, 80*, 267–276.

Zárate, G., Perez Chaia, A., González, S. & Oliver, G. (2000b). Viability and β-galactosidase activity of dairy propionibacteria subjected to digestion by artificial gastric and intestinal fluids. *Journal of Food Protection*, *63*, 1214-1221.

Zárate, G. (2001). Propionibacterias como Probióticos en el Metabolismo Intestinal de Lactosa. *PhD Thesis,* Universidad Nacional de Tucumán.

Zárate, G., Perez Chaia, A. & Oliver, G. (2002). Some characteristics of practical relevance of the β-galactosidase from potential probiotic strains of *Propionibacterium acidipropionici*. *Anaerobe*, *8*, 259-267.

Zárate, G. (2012). Dairy Propionibacteria: Less Conventional Probiotics to Improve the Human and Animal Health. In *"Probiotics in animals"*, ISBN 978-953-51-0777-4, Chapter 8: 153-202. Intech Open Access Publisher.

Zárate, G. & Perez Chaia, A. (2012). Influence of lactose and lactate on growth and β-galactosidase activity of potential probiotic Propionibacterium acidipropionici. *Anaerobe*, *18*(1), 25-30.

Zhang, W., Wang, C., Huang, C. Y., Yu, Q., Liu, H., Zhang, C. & Pei, X. (2011). Construction and expression of food-grade β-galactosidase gene in Lactococcus lactis. *Current Microbiology*, *62*, 639–644.

Zhong, Y., Huang, C. Y., He, T. & Harmsen, H. M. (2006). Effect of probiotics and yogurt on colonic microflora in subjects with lactose intolerance. *Wei Sheng Yan Jiu – Journal of Hygienic Research*, *35*, 587–591.

In: Lactose
Editors: David Green and Emma Lee
ISBN: 978-1-62808-198-5

Chapter 2

Milk, Lactose, Lactase: The Medical Adventure

Andrew Szilagyi[*]
Division of Gastroenterology, Department of Medicine
Jewish General Hospital, McGill University School of Medicine,
Montreal, Quebec, Canada

Abstract

Lactose is a disaccharide which is the main carbohydrate present in milk of most mammals, including humans. At birth mammals possess lactase for digestion but lose this enzyme after weaning. In humans however, about 1/4 to 1/3 of the world's population retain lactase throughout adulthood, allowing continued consumption of milk.

The consequences of this human divide are the incurring of digestive symptoms in those who have limited intestinal lactase and consume large enough quantities at a time.

The medical issues arising from this dichotomy then include the consequences of lactose associated symptoms and the more important consequences of diary food avoidance. In addition dairy foods are linked or hypothesized to increase or decrease risks for many diseases. Perhaps one of the most interesting aspects of the world lactase dichotomy is the relationship of population distribution and geographic patterns of modern

[*] Address 3755 Cote St Catherine Rd, Rm E110 H3T 1E2 Montreal Que. Tel 514 340 8144 ; Fax 514 340 8282 ; E-mail : aszilagy@jgh.mcgill.ca.

day, so called "western" lifestyle linked diseases. The asymmetric distribution and increasing incidence/prevalence of these diseases in developing nations sheds light on the possible co evolutionary impact of lactase pheno or genotype on such diseases. The development of the asymmetric lactase patterns and consequent medical aspects will be discussed.

Introduction

Lactose is a disaccharide which is the main carbohydrate present in milk of most mammals, including humans. It consists of two hexoses, glucose and galactose (Figure 1).Virtually all newborn mammals are born with the intestinal brush border enzyme lactase phlorizin hydrolase (LCT, or LPH), which hydrolyzes lactose into its components. These are then absorbed through the intestine. However, newborn lose this enzyme after weaning (lactase non persistence, LNP). In some mammals lactase may be induced by reintroduction of lactose into the diet, but humans are different. In about 1/4 to 1/3 of the world's population enzyme levels are retained throughout adulthood (lactase persistence, LP). As such this is a dominant Mendelian trait and allows continued consumption of dairy foods (including high lactose containing fluid milk) throughout adulthood. In the rest of the population with LNP, enzyme LPH is reduced in intestinal tissue and is not inducible by reintroducing lactose into the diet [1].

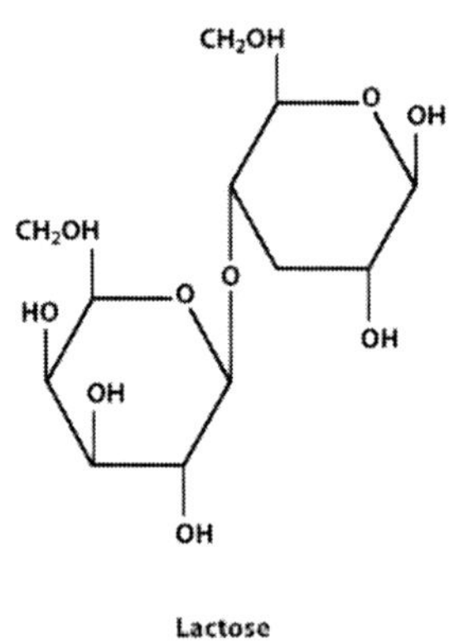

Figure 1. Chemical Structure of the disaccharide lactose consisting of 2 aldohexoses: glucose and D – galactose.

This chapter will outline the evolution and spread of the dominant genetic trait responsible for continued adult digestion of lactose. The genetics involved will be described. Secondary causes of lactose maldigestion will be outlined, but the bulk of the chapter will discuss the medical relevance which divides humanity into those who can and those who cannot digest the sugar. The diagnostic methods used to distinguish LP form LNP will be outlined. A discussion of the effect of this divide on other diseases and modern geographic distributions will be addressed.

The Evolution and Spread of Lactose Digestion

In order to understand the current distributions of LP/LNP, migrations of early humans and the introduction of animal husbandry and milking practices in culture need to be elaborated. The evolution of lactase persistence and adult ability to digest lactose is a relatively recent event in human history and is believed to still be under selective pressure [2]. The dominant trait is thought to have evolved around 7.5 $x10^3 - 10^4$ years ago [3]. Earlier than this, humans like other mammals were thought to have no adult ability to digest this disaccharide. A popular website outlines over 10^5 years of human migration (http://www.bradshawfoundation.com/journey/). People of central Asian origin are thought to have travelled over an existing land mass connecting Asia and North America some 18 $X10^4$ years ago during the last Ice Age. From here, territories of modern Canada, United States and Central and South America are thought to have been populated. Comparisons of Y chromosomes between native Americans and Siberians support this hypothesis [4]. Migrations into the south pacific may have occurred more directly from Africa [5. Similarly in Europe early Neolithic man was unable to digest lactose [6]. Then in the last ten thousand years lactose digestion became possible. The explanation for this evolutionary step is now strongly supported by evidence of the introduction of animal husbandry and subsequent milk drinking occurring almost simultaneously in several parts of the old world [3]. This hypothesis is referred to as the gene-culture co evolution or gcc hypothesis. This hypothesis proposed first by Simoons [7, 8] and then Holden and Mace [9] ultimately prevails over two other utilitarian hypotheses. An earlier hypothesis was put forward by Flatz and Rothhauwe to account for a close relationship between a north to south gradient of increasing population frequency of LNP status in Europe [10]. They postulated that lactose digestion dominance preceded introduction of herding. The development of LP status occurred as a compensation for low population exposure to sunshine and skin synthesis of vitamin D. The ability to digest lactose would allow milk drinking with exposure to large amounts of calcium. In turn this reduces risk of rickets an important cause of poor bone formation. This is referred to as the calcium assimilation hypothesis. Another hypothesis by Vullo and Anderson postulated that LNP status protects against malaria [11]. There is little support for this hypothesis and is discussed below.

Most genetic anthropological studies and simulated models favor the gcc hypothesis, but in Europe the latitudinal pattern is still not fully explained [12, 13]. According to the prevalent gcc hypothesis, milk drinking originated in the central Balkans and populations from here migrated north and then east [12, 13]. Indeed several reports favor such migrations into central Russia and subsequently into northern India and possibly northern China [14-16]. At around a similar time frame in the Middle East and various tribes in Africa also domesticated camels and other cattle resulting in convergent evolution of different genetic polymorphisms controlling LPH and ability to digest lactose [3, 17]. Then after the rediscovery of the new world by Columbus, migration of north western and south western Europeans to the new worlds both in North/South America and Australia/New Zealand introduced LP populations in large numbers.

Genetics of Lactase

The gene for lactase has been located on Chromosome 2 [18]. With the exception of congenital lactase deficiency the normal regulation of LCT controlled via transcription from a promoter region located upstream from the 5' region of LCT, 14kb pairs in a cis position. Such cis position regulation leads to an imbalance of alleles and is characterized by allelic instability. Cis acting genes generally have promoter, enhancer, and silencer functions or have an insulating effect on target genes. These cis acting genes facilitate adaptation to soft environmental sweeps [19]. Even prior to identification of specific single nucleotide (SNP) polymorphisms it was recognized that several haplotypes (A haplotype in genetics is a combination of alleles (DNA sequences) at adjacent locations (loci) on the chromosome that are transmitted together) are in evolved with adult LP status [20] In 2002 Ennatah et al identified a locus (chromosome 2q21-22) within an intron of a nearby gene;

minichromosome management 6 (MCM6), 13910 bp downstream from LCT [21]. This SNP correlated 100% with LP status. In this paradigm substitution of Thymidine for Cytosine rendered the individual LP, such that T/T is homozygous dominant LP, T/C is heterozygous LP but on intestinal biopsy enzyme of lactase was intermediate and C/C is homozygous LNP with <10units (u) lactase/g of tissue [22]. Enattah et al also described a second SNP in Europeans, the G/A -22018 but does not seem to have a promoter control function in Europeans. This SNP may be responsible for LP status in Northern Chinese [16] in addition to other polymorphisms.

Table 1. The various polymorphisms described for lactase persistence are shown

Polymorphism	C/T 13910	G/A22018	C/G13907	G/C14010	T/G13915
Described by	Ennatah[21]	Ennatah[21] Xu [16]	Tishkoff[3]	Tishkoff[3]	Tishkoff[3] Imtiaz [24]
Populations affected	N Europe Ancestors in NA*,SA* Aust* Central Asia N India	Similar to C/T13910 N China	African Tribes (eg Beja in Sudan)	African Tribes (eg Kenya And Tanzania)	African Tribes (eg Kenya) Middle East (eg Saudi Arabia)

* NA- North America, SA – South America, Aust – Australia.

These SNP changes appear to have developed around 5000 – 10000 years ago in different locals and have been termed as examples of convergent evolution [3]. Lactase persistent dominant polymorphisms are likely to expand as additional populations are evaluated. As well, it is hypothesized that there is ongoing evolutionary pressures on these polymorphisms to this day [2].

Since different haplotypes have been recognized to be responsible for different SNPs the search for dominant genes also expanded to Africa and the Middle East [17, 23]. There populations (tribes) whose ancestors domesticated animals for milking generally were lactose digesters. Indeed 3 new different polymorphisms were described close to the European polymorphism. These included C/G-13907, T/G-13915, and G/C-14010 [3]. Of these the T/G -13915 SNP has been reported to be the principal polymorphism in Saudis [24]. Based on simulation models it is predicted that other polymorphisms will be found as different populations are systematically evaluated [25]. However, there is evidence that the leading LP polymorphisms in central Asia and northern India have so far retained the importance of the C/T – 13910 SNP [14, 15]. These findings support the contention that these world regions are populated by migrants from northern Europe. The currently discovered genetic of dominant LP status are listed in Table 1.

Symptoms Related to Lactose Consumption

If one "Googles" the title of this section there will be 1,330,000 sites displayed in 0.36 seconds. Such a volume attests to the interest and importance of this topic to the general public. However, there has been and there continues to be confusion about such symptoms and the meaning of different terms related to them. The main, often interchanged terms to denote symptoms are outlined in Table 2. It should be noted however, that each term has different meanings. For example lactase deficiency (LD) refers to a reduction in quantity/g of LPH in intestinal tissue for any reason. This LD in turn leads to the inability to digest lactose and is labeled as maldigestion (LM). This could be due to the rare congenital absence of LPH [26], a rare variant with non functional enzyme from birth described in a Japanese infant [27], secondary causes such as diseases which lead to villus small bowel damage (celiac disease, giardiasis, rotavirus infections, surgical resection, drugs and many others) and common primary adult LPH deficiency (lactase deficiency, LD) which leads to LNP status. With exception of serious consequences to nutritional deficiency in congenital absence or premature birth, the primary symptoms of LPH loss in adults are gastrointestinal. These include gas, abdominal bloating, cramps, and less often diarrhea and vomiting. These symptoms are termed lactose intolerance (LI). They are not specific to lactose and may occur with any food intolerance (especially carbohydrates from mono, other di, oligosaccharides and polyols like sorbitol). In addition other foods like coffee, fats and caffeine in other food scould reproduce some of the symptoms. Furthermore, diseases affecting the small or large bowel and psychological stressors could also aggravate such symptoms. More recently the term lactose sensitivity (LS) was coined by Mathews [28] and is also applied to the Inflammatory Bowel diseases (IBD). This LS refers to additional systemic symptoms and general fatigue, headaches and even pruritus [29]. The problem with these different terms is that they are not interchangeable and the relationship to lactose specifically may be lost as descriptions and tests of LM are deleted. Yet often these terms are used interchangeably. Partly because of these different terms and because LI is often defined merely by self reported beliefs, there is no clear knowledge of the prevalence of LI in large Western societies like the USA and Canada. Furthermore development of LI in LNP persons can be modified by ingestion of lactose with different foods of different energy density and by low doses ingested more frequently. For example, lactose in water will precipitate symptoms more readily than lactose in high fat milk or chocolate milk [30].

The explanation for the development of symptoms with LM relates to the quantity of lactose that escapes digestion in the small bowel where the bulk of LPH is found (jejunum > duodenum>> ileum). It was pointed out above that in adult LNP at least some enzyme remains and that depending on dose, manner in which lactose is consumed and some physiologic conditions symptoms related to LM are modifiable. There are three non exclusive mechanisms by which symptoms could develop. In the first instance a very large dose of a lactose challenge in aqueous medium(like water), for example, 50 -100 mg (the lactose content of 1 - 2L of milk) will rapidly overwhelm whatever residual lactase is present in an LNP individual. As a result intestinal transit time will decrease. The residual lactose will rapidly reach the colon in large amounts and exert osmotic forces which will induce water to passively enter the colon. This will result in cramps, diarrhea and even on occasions vomiting. However, people who consider themselves to be LI will not likely consume such high levels in one session.

In the second scenario more modest doses of lactose in the range of 20-25 g (the content of about 2 cups of fluid milk) can also induce symptoms in LNP individuals. While the conditions mentioned in the first instance may also be at play here, intestinal (mainly colonic) microflora has been hypothesized to play an important role. In this scenario undigested lactose is metabolized by microbes which convert lactose into gases (hydrogen, carbon dioxide and methane) and the short chain fatty acid (SCFA) lactate and acetate [31]. In turn these substrates (especially lactate) stimulate other bacteria which can produce other SCFAs (propionate and butyrate) [32]. He et al postulate that in naïve LNP subjects it is an ab- normal production of SCFAs which lead to the development of symptoms [31]. In LNP persons who continue to consume lactose a microflora change occurs with the expansion of so called lactic acid producing bacteria. These include lactobacilli [33, 34] and bifidobacteria [35-37]. The expansion of these bacteria successfully compete for preferential metabolism of lactose [38] and are subsequently associated with less gas production and perhaps reduction of symptom [39]. This process is called colonic adaptation and will be discussed further below.

In the third scenario symptoms are supposed to occur at low levels of lactose ingestion (< 15g of lactose). In practice symptoms incurred at this level are independent of LM [40-42]. In the case of LM subjects the development of symptoms will depend on the amount of lactose digested by residual intestinal LPH. There are a number of studies dealing with the minimal amount of lactose that can be digested by LNP subjects in a single session. In general the doses range from 6g to just over 10g of lactose [42-44]. In keeping with this notion a study examining symptoms incurred by ingestion of < 1g lactose , an amount which may be used in the production of medications did not find any problem at this dose [45]. In the usual scenario of symptoms incurred at these doses several studies examined in a blinded fashion the ability of persons believing themselves to be severely LI to distinguish between single ingestion of lactose free or regular lactose containing milk. In the study of Suarez et al such people were unable to distinguish lactose free from regular milk [46]. Later, a second study was published by the same group demonstrating the same phenomenon with 2 cups of milk separated in time [47].

Table 2. Terms related to lactose and its digestion are shown. The terms should be distinct

Lactase Persistent LP	The dominant genetic trait in adults with continued ability to digets lactose
Lactase Non Persitent LNP	The natural decline in intestinal lactase to < 10u/g of tissue which leaves adults with minimal ability to digest lactose
Lactase Deficiency LD	Reduction of intestinal lactase enzyme from any Cause either genetic (LNP) or any secondary causes like diseases of the proximal small bowel mucosa
Lactose Maldigestion LM	Inability to digest lactose for any cause primary (LNP) or secondary causes
Lactose Intolerance LI	Symptoms resulting from the ingestion of lactose Including flatus, gas, bloating, cramps, diarrhea and rarely vomiting. LI may occur without LM
Lactose Sensitivity	Symptoms with or without symptoms of LI and systemic features depression, headache fatigue These features need to be further clarified and May occur with or without LM

A meta –analysis of symptoms by self reported LI subjects demonstrated that there is a significant over statement of symptoms in this group [48]. Similarly Cassellas et al reported a significant discrepancy between symptoms reported for LI by subjects who subsequently underwent testing for LM [41]. Another group reported that rather than LM status, somatization as found on psychologic testing was better correlated with self reported LI. Furthermore these persons also scored less well on measurements of quality of life. [40]

It is therefore noted that there is much variability in symptoms related to lactose ingestion. In general, symptoms tend to be worse in LM persons but LI symptoms are not limited to LM (LNP) subjects. The presence of symptoms in LP (as well as in some LNP) persons is governed by other parameters which may be related to additional psychological factors as well. As a result, self-reported LI is less specific to LM and the diagnosis of maldigestion is not made on history alone. However, under such uninvestigated conditions lactose induced symptoms are likely not specific and could apply to other offending foods as well. It should be also noted that lactose maldigestion is not more frequent in irritable bowel syndrome [49,50].

Tests for Lactose Maldigestion

Adult LM leaves about <10u/g of lactase on the intestinal brush border. Direct measurement of this enzyme in biopsies could be the ideal test for diagnosing LM (of any type). However, this test is too invasive and not suited for population or individual clinical diagnosis of LM. Nevertheless, biopsies are not perfect because the decline in lactase in adults with LM may be spotty and so biopsies could be falsely negative [51]. Of note however, is that in the last 5 years the intestinal biopsy test for LPH deficiency has returned in the form of a “Quick [R] “ test which allows rapid diagnosis of LPH deficiency during gastroscopy [52]. Again however, the test is suited to special situations (in particular it allows the exclusion of secondary causes of LM and LD), and is difficult to implement for widespread use. Instead prior to the description of genetic tests, indirect tests of lactose maldigestion were introduced. Although several tests were available, only two are still used clinically [53]. The more widespread test used is the Lactose Breath Hydrogen (LBHT) and occasionally the older lactose tolerance test (LTT) can be still used.These indirect tests have been correlated with intestinal biopsies.

Both tests have limitations and it should be clear that the tests are mirror “images” of each other. The LTT measures the host’s ability to breakdown lactose into its monosaccharide components. Blood glucose is then measured in a fixed time interval (usually 2 hrs) after a baseline is measured. A lactose load is given and a rise above 1.-1.4 g/L is considered a positive test and establishes LP status [54]. A rise less than this, is a negative outcome supporting LNP status. On the other hand the breath test is interpreted in the opposite way. A lactose load is given and the amount not digested by the host reaches the lower intestine. Here bacteria metabolize lactose and form gases and short chain fatty acids as discussed above. Hydrogen is a key element that diffuses across the colonic membrane, enters the circulation, traverses the lung where along with some other gases like carbon dioxide (and in a significant number, methane) are exchanged for oxygen. The exhaled hydrogen is measured, either by a electronic hydrogen sensor or as samples drawn from an airtight bag into which participants exhale at timed intervals. Samples are injected into a compact gas chromatograph and hydrogen concentration is measured. Thus the positive rise in hydrogen by 20 parts per million (ppm) above the baseline denotes a person with LNP status. A failure to raise hydrogen levels usually means LP status [53,55].

Both tests have limitations [56]. However, because the breath test is actually a bacterial function test dependent on the host, several caveats need to be mentioned. Firstly about 5-15 % of the population may not have bacteria which produce hydrogen. Although, if the limit of rise of hydrogen is lowered to 10ppm the likelihood of sensitivity is increased although specificity may be lowered [57]. Secondly an ethnically indecent portion of the population harbor methanogenic bacteria. These bacteria utilize hydrogen gas produced to form methane (CH_4), thus reducing the response to lactose. As a result some authors recommend including methane measurement to increase sensitivity [58]. The third variable which affects breath tests (but not yet studied and is not expected to affect LTT) is the notion of bacterial adaptation to lactose [39]. In this paradigm continued lactose consumption in LNP persons leads to altered bacterial flora. These include lactobacilli and bifidobacteria which produce less or very little hydrogen, possibly rendering the test less sensitive. For these and other reasons the correlation between the LBHT and LTT is modest to moderate at best [59].

Most recently after the description of the North European C/T -13910 lactase promoter polymorphism RIA tests became available to genotype persons. This test however, is more difficult to use for daily clinical evaluation because the tests do not measure whether a person is LI. Even the notion of LM is age dependent so these tests are more relevant to epidemiological and anthropological genetic studies. The genetic tests have also been well correlated with intestinal enzyme levels [22]. In addition our group evaluated in a meta-analysis the accuracy of the two most frequently used indirect tests discussed above to predict genotype within the C/T-13910 polymorphism [60]. A literature search revealed 19 studies from worldwide centers of which 17 discussed the LBHT, and 2 the LTT uniquely. In 3 of 17 LBHT studies both tests were evaluated. The overall sensitivity of the LBHT was 0.88 (CI, 0.85 – 0.90) and specificity was 0.85 (CI, 0.82 – 0.87). In sub-analyses, use of a 50g lactose load and the failure to include children (< 18 yrs), increased sensitivity and minimally changed specificity (lower with the high lactose load, and higher in the omission of children). Despite the poor correlation between the LBHT and LTT, the latter also performed well against the genetic tests albeit with very few studies (5 publications). The sensitivity was 0.94 (CI, 0.90 – 0.97) and specificity was 0.9 (CI, 0.84 – 0.95). This latter analysis was of strategic importance in order to allow comparisons of various disease rates to be correlated with population distributions of LP/LNP (to be discussed below). The study supports the contention that indirect tests individually do reflect genetic traits as long as the specific traits are measured for appropriate populations which may have different polymorphisms from the north European type.

Epidemiology of Dairy Food Consumption, Lactase Phenotypes and Diseases

In the last few decades, two variables have been recognized that impact on geographic distributions on modern diseases. The first is the constant relationship of the position of the earth with respect to the sun. These polar/ equatorial directions are demarcated by latitudes in northern and southern hemispheres, diminishing toward the equator. Lower latitudes are also associated with increasing regional sunshine exposure. Sunshine is divided into visible light (400 – 700 nm) and ultraviolet light (UV). Of these UVC never reaches earth. UVA (320 - 400 nm) may be linked with melanoma [61] and UVB (280 – 315 nm) is responsible for skin synthesis of vitamin D from 7 – dehydrocholesterol [62].

With the discovery in the last two decades of important non skeletal functions for vitamin D3, the relationships between many diseases have been elaborated. A large number of diseases have been identified with modern day " western" cultures and lifestyles. These include at least 14 cancers, leukemia, multiple sclerosis and the inflammatory bowel diseases, [IBD; Crohn's disease and ulcerative colitis] [63 – 68]. In general all these diseases tend to follow diminishing rates from north toward the equator. However, these reductions in rates also occur eastward (if the reference points are highest in north America, western Europe toward Asia).Disease rates in both latitudinal and non latitudinal directions are increasing with progressive "westernization" and/or industrialization.

It is observed, that especially in Europe there is a close relationship between latitude, UVB and LP/LNP distributions. These relationships were evaluated with both genetic models of LP [25] as well as phenotypic distributions derived from older observations [69] with concurrence between the outcomes. These relationships are much weaker outside of Europe. However, to date, national disease rates on a global level also show relationships to these three variables. The direction of change with LP/LNP distributions in disease rates seem to parallel directions achieved with latitude and UVB. That is decreasing disease rates with either increasing LNP fractions of populations or decreasing latitude associated with increasing UVB exposure.

However, because the leading hypothesis for emergence of LP is association with herding [3,7,9], LP/LNP distributions are thought to depend largely on gcc and population migrations. The independent development of LP status from latitude /UVB exposure suggests that disease patterns likely depend on interaction of both variables [70].

For some diseases like colon [71] and ovarian cancers [72] as well as the IBDs [73] it was recognized almost 2-4 decades ago that disease rates diminished with increasing rates of population LNP status. In a more encompassing study evaluating the relationship between the rates of seven "western" diseases and population LP/LNP distributions, two independent effects were noted [74]. The first effect on diseases was defined by the relationship of national per capita dairy food (DF) consumption (Fig 2). Relationships were more varied with DF consumption and the epidemiological data were also compared for outcome with existing meta-analyses. The second relationship was the effect of population distributions of LP/LNP, where LNP was inversely and significantly associated in each of the seven "western" diseases and directly with the one "eastern" disease (Table 3).

In this paradigm between dairy food, consumption and lactase distributions, six hypothetical non mutually exclusive interactions are possible; I. Dairy food and lactose dependent: a. risk decreased, b. risk increased, II. Dairy food dependent and lactose independent: a. risk decreased, b. risk increased, III. Dairy food independent (? Co – marker effect): a. risk decreased and b. risk increased. These relationships are discussed below and displayed in Table 4.

Table 3. Calculated Odds Ratios (OR) for Each Disease Either by per Capita Dairy Consumption or National Lactase Non Persistance Prevalence (LNP) Rates Using Negative Binomial Regression Reprinted with permission from Shrier et al Nutr Cancer 2008 [ref 74]

	Dairy Consumption			LNP Status		
	Coefficient (SE)	OR per 10	g/kg (CI)	Coefficient (SE)	OR per 10%	LNP (CI)
Incidence						
Crohn's	.0079 (.0046)	1.08 (0.98–1.18)	P=0.09	–.018 (.0042)	0.83 (0.76–0.90)	P<0.01
UC	.0081 (.0025)	1.08 (1.03–1.13)	P<0.01	–.0087 (.003)	0.91 (0.86–0.97)	P<0.01
Mortality						
Prostate	.0036 (.0016)	1.03 (1.00–1.06)	P=0.03	–.0115 (.0021)	0.89 (0.85–0.92)	P<0.01
Ovarian	.0022 (.0014)	1.02 (0.99–1.05)	P = 0.10	–.0075 (.0019)	0.92 (0.89–0.96)	P<0.01
Colorectal	.0041 (.0015)	1.04 (1.01–1.07)	P <0.01	–.0081 (.0022)	0.92 (0.88–0.96)	P<0.01
Lung	.0028 (.0014)	1.02 (0.99–1.05)	P = 0.06	–.0062 (.0023)	0.93 (0.89–0.98)	P<0.01
Stomach	–.0043 (.0015)	0.95 (0.92–0.98)	P<0.01	.0115 (.0027)	1.12 (1.06–1.18)	P<0.01
Breast	.0018 (.0016)	1.01 (0.98–1.05)	P=0.29	–.007 (.0026)	0.93 (0.88–0.98)	P<0.01

[a] The coefficient represents the increase in log risk per 1 unit increase of the variable. Therefore, the OR per 10% increase in LNP is equal to "exp(coefficient^{10})," and the OR per 50% increase in LNP is equal to "exp(coefficient^{50})."

Abbreviations are as follows: CI, confidence interval; UC, ulcerative colitis.

Table 4.

A. Diseases linked with relationships to either dairy food consumption or disease rate change with lactase proportion distributions* See text for discussion of diseases

Dairy Food Causes	Dairy Food Protects	Dairy Food Independent
Prostate Cancer	Colorectal Cancer	Breast Cancer
Ovarian Cancer	Stomach Cancer	Lung Cancer
	Osteoporosis	Crohn's disease
	Bone Fractures	Ulcerative colitis

B. Diseases evaluated or not evaluated to be related to Lactase distributions (inverse with Lactase non persistence but have relations with Ultraviolet light (UVB). Potential mimicry of above

Cause	Protective	Independent
Cataracts		Multiple Sclerosis
Testicular Cancer	Hypertension	Lymphoma
	Diabetes II	Leukemia

* The arbitrary divisions are based on best current information and may not be mutually exclusive.

Specific Genes Evaluated in Relationship to Lactase Distributions

The evolution of lactase persistence in the Neolithic period allowed LP persons to consume fluid milk, increasing their nutritional intake. However, mortality associated with enteric pathogens through diarrhea increased. It is hypothesized that some genes co-evolved with LP in order to counter act mortality caused by enteric pathogens. To date all genes evaluated in relationship to milk drinking originated from Europe.

Genes thought to have evolved for protection against diarrhea include the Cystic fibrosis CFTR gene, A gene for Oxytocin receptor, HLA polymorphisms and NOD2/CARD15 microbial sensor system.

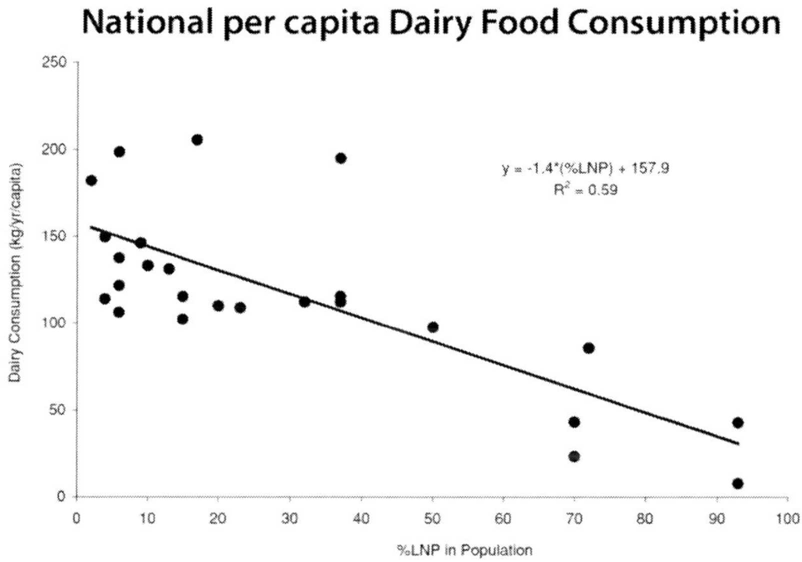

Figure 2. Dairy food consumption for each country is plotted against the percentage of the population that are lactase non persistent (LNP). Because LNP subjects experience side effects due to lactose if they are not adapted, they consume less dairy foods. Reprinted from Shrier et al Nutr Cancer 2008 [74] with permission.

Cystic Fibrosis and Mutations in the CFTR Gene

The gene responsible for cystic fibrosis(CF) have been postulated to be more common in LP populations [75,76]. Cystic fibrosis is considered to be the most deadly gene affecting Europeans and their descendents. The autosomal recessive trait results in a loss of function in the Cystic Fibrosis Transmembrane conductance Regulator of Cl- channel and water excretion. It is also hypothesized that this gene in its heterozygous form counteracted diarrhea induced by milk drinking before the evolution of lactase persistence [76]. An evaluation however, of lactase mutations in patients with CF showed a higher frequency of the C allele of the European C/T-13910 polymorphism in homozygotes of severe CFTR mutations [77]. In cases of mild CFTR mutations there was a lower frequency of the C allele. The authors concluded that LNP was equally distributed between controls and patients. The expectation would be an increased rate of LP in CF patients since the ancestral risk was in milk drinkers. The fact that more severe cases had a higher rate may still link the evolution of these two genes. However, as noted earlier latitude and lactase are closely linked in Europe, making distinction of evolutionary pressures and interpretation complex.

Non Parturition Effects of Oxytocin

Recently the hormone oxytocin has been described as being important in pain modulation and gut, especially colonic motility. Genetic modification may play a role in response to enteric pathogens. To determine whether genetics of oxytocin (OXT) may be involved in pain modulation in IBS a study using DNA from 131 well characterized IBS patients 408 healthy controls and 299 asymptomatic Swedish blood donors were compared. As well the genotypes of CC and TT were simultaneously evaluated. There are 2 chromosomes involved in OXT control. One is a promoter and the other is an OXT receptor on different chromosomes both removed from the chromosome for lactase. It was found that an SNP in the promoter region of OXT was more frequent in LNP (CC) persons but did not differentiate patients with IBS [78]. Although neither of these polymorphisms are linked with the lactase gene, nevertheless are unequally distributed between lactase phenotypes.

Human Leucocyte Antigens and Immunity

A third group of polymorphisms associated with resistance to pathogens are the HLA

(Human Leukocyte Antigens) system part of the Major Histocompatibility Complex located on chromosome 6 (6p21.31). A large proportion of genes are involved with the immune system and have a very high rate of heterozygosity (>80%). In an analysis of a large number of alleles from Europe a spatial pattern was observed in about 1 % but mostly with a latitudinal gradient. The authors concluded that selective pressures during the Neolithic period, most likely new epidemic diseases were responsible. A parallel emergence of herding, lactase persistence, and milk drinking with subsequent exposure to enteric pathogens could account for these patterns [75].

Elimination of Milk Transmitted Pathogens NOD2/CARD15

A fourth genetically determined system, caspase-activation and recruitment domain gene 15 (part of the also called Nod 2 system) located on chromosome 16q12 was the first gene described to increase risk of Crohn's disease (CD). Three mutations increase the risk of CD about 40 fold if homozygous and about 4 fold if heterozygous [79]. The mutations SNP8 (containing R702W), SNP 12 (containing G908R) and SNP 13 (containing 1007fs) are responsible for 80 % of the risk and these are largely confined to

Caucasians from Europe and their decedents in the new worlds [80]. The NOD2 system through interactions with bacterial particle sensing Toll receptors [and other genes with autophagy function: IRGM, and ATG16L1, the latter located on chromosome 2q37 (the same chromosome as lactase)] are responsible for excluding bacteria from traversing the intestinal membrane and invading the human host [80, 81]. As with NOD2, A mutation at T300A confers an increased risk for CD [81]. The development of this system is hypothesized to have developed in response to the practice of milk drinking [82]. Specifically common bacteria which are passed from cattle through unpasteurized milk; Brucella, Listeria and Mycobacterium Bovis are targets of this system.

Latitudinal Changes of Taster / Non Taster Status

Another polymorphism evaluated for co-evolution with lactase is the gene for taster status. Persons with the recessive gene for non taster status may consume more cruciferous vegetables because the associated bitter taste is reduced. The suggestion is that these vegetables may protect against cancer. Sacerdote et al evaluated taster status in northern against southern Italians and related this to LP/LNP status in the same population. They reported that non taster status was somewhat more common in the northern population where LP status was also higher than in the south [83]. The relationship of this polymorphism however, may be coincidental.

Disease related to Dairy Food Consumption and the Lactase Dichotomy

Dairy foods contain a significant array of nutrients which could impact on disease modification [84].Although, calcium, fat, potassium, magnesium and vitamin D (added in North America) content are important elements in this function, lactose is also a key element toward disease modification. The lactose content of dairy foods has an important impact on quantities consumed by LNP as well as a prebiotic effect limited to LNP subjects [74]. These two qualities should affect any disease hypothesized to be related to dairy food consumption.

Disease Related to Evolutionary Development of the Lactase Dichotomy

Evolution of lactase genotypes that resulted in improved nutrition in LP persons likely had additional impact on the evolution of other genotypes. Genes which conferred an advantage to ancestors may behave differently in a modern environment. The question of co evolution of other genes then poses the question whether genes evolved with LP or was independent of lactase but included those prevailing environmental pressures. These latter changes could be related to latitudinal effects as has been suggested for loss of pigmentation [85] and sodium homeostasis and hypertension [86,87].

In the next sections disorders linked with DFs consumption or lack thereof will be discussed. The classification outlined above for DF consumption dependency or independency is followed.

Dairy Food Dependent: Disease Risks Reduced by Lactose/Dairy Foods

A recent 2010 NIH consensus conference on lactose intolerance and health concluded that the main adverse outcome of LI is the avoidance of milk and other dairy foods (DFs) [88]. The independent panel noted that LI without proof could cross generation lines because of restriction of DFs in children of parents who believe themselves to be LI and thus believe that symptoms will develop in their offspring. Those avoiding DFs were found to be deficient in some nutrients, especially calcium and vitamin D [89].

The main effect of DFs avoidance is the possibility of osteopenia and osteoporosis. These then could lead to higher rates of bone fractures. At the 2010 NIH conference, Wilt et al reviewed the literature evidence for the effect of different categories of lactose intake (as a surrogate for dairy food intake) on bone mineral content (BMC), bone mineral density (BMD) and fracture risks [90].The analysis included four groups; general population, vegans (strict vegetarians), persons with self reported LI and or LM, and proven LNP persons. Fifty –two studies, including 9 randomized controlled trials were included, which dealt with lactose intake and bone health. The authors concluded that there was low level evidence, based on observational studies that milk avoidance may be associated with osteopenia, BMC, BMD and increased risk of bone fractures. They noted that in adults with LI, LM, the CC genotype (LNP) or reduced lactose intake, may be associated with increased bone fractures. However, this was reported only from a single Finnish and a study dealing with vegans. Women were found to be at higher risk. There was also better evidence that increased lactose intake did improve BMC and BMD. A more recent report from two Dutch centers, not included at the NIH conference, found that while the CC genotype was associated with lower calcium intake and lower serum ionized calcium, BMD or fracture risk were unaffected in these elderly people [91]. In general however, the panel interpreting the presented data suggested that more studies are needed on the relationship between bone health, lactose intolerance and dairy food avoidance [88].

Other medical consequences relate to the metabolic syndrome. This includes obesity, hypertension, hyperlipidemia and type 2 diabetes mellitus. A large study including over 1800 Australian subjects evaluated 145 dietary items including regular fat dairy foods consumed over 10 years. The authors reported a 59% lower risk of metabolic syndrome in the highest quartile of regular fat DFs intake. Interestingly no association between total DFs intake obesity and diabetes was noted [92]. In an earlier, similarly large, long term study of subjects with self reported LI and DFs avoidance Nicklas et al reported an inverse association between total DFs and metabolic syndrome as well as diabetes [93].

The effect of DFs consumption specifically on obesity is controversial. A recent meta-analysis of 29 randomized controlled trials found only a modest benefit of DFs consumption on limiting obesity in the short term (94). In the absence of energy intake restriction no long term benefits to DFs consumption were noted.

The relationship with obesity and metabolic syndrome is further confused by a multicenter European study where a large number, over 31 thousand subjects were evaluated for lactase genotype and basal metabolic index (BMI). The report clearly linked an increased BMI to the lactase persistence status. The study did not evaluate DFs consumption in the test population [95]. However, as will be discussed below this variable is important because the relationship observed may be a false impression as is clearly the case with colorectal cancer (see below).

In conjunction with DFs restriction there is also a risk of inadequate vitamin D intake. The reason is that in the US and Canada milk is fortified with this vitamin. Vitamin D deficiency has been linked with an increased risk and an increased likelihood of worse outcome with a large number of diseases. Among these are at least 14 cancers [96,97], Lymphoma [98], Leukemia [99],

Inflammatory Bowel Diseases [100] and multiple sclerosis [101]. In the absence of vitamin D supplements therefore milk is the largest source of dietary intake of vitamin D. Ethnic groups such as African American and Hispanic groups may be at a higher risk for vitamin D deficiency also because of reduced skin production and perceived LI [102].

The other major area where dairy foods exert protective effects is on gastrointestinal cancers [103]. It is emphasized that in epidemiological studies countries with the highest national per capita DFs consumption have the highest rates of colorectal cancer. Paradoxically, DFs, calcium and vitamin D all exert protection against colon cancer [74,104]. The paradoxical relationship between a disease epidemiological link to a putative aggravating agent(s) and the actual findings from patient based studies is known as an ecological fallacy (Definition at: http://michaeljohnsonphilosophy.com/wp-content /uploads/2012/10/ecological-fallacy.pdf). A paradigm of false epidemiological connection may also apply to obesity as outlined above.

The primary mechanism of reduced cancer is the antiproliferative effect of calcium on aberrant foci and later advanced polyp formation. Initial reports on the effects of calcium suggested that at least 1200 g of calcium is needed for this effect [105,106]. However, the addition of vitamin D which has important antineoplastic cell signaling function may reduce the need for the level of calcium since the two may have additive effects [107]. However vitamin D added to milk is limited to western countries. It is of interest that DFs have been reported to protect against CRC in regions of the world with either relatively high frequency of LP persons (such as north western Europe, North America and Australia where large amounts of DFs are consumed) and high frequency of LNP (such as Japan and China where DFs consumption is about a third of western intakes). Protective effects in countries with more evenly distributed LP and LNP (such as southern Europe and South America) populations, DFs appear to have less of an effect [108].

The explanation for protection in LNP populations which consume less DFs is unclear but a prebiotic effect of lactose in such populations could substitute and compensate for lower calcium intake [37,109]. Prebiotics promote the preferred growth of probiotic type bacteria (lactobacilli and bofidobacteria). The anti-carcinogenic effects of such probiotic bacteria have been reported [110]. It should be noted that with lactulose (a disaccharide consisting of galactose and fructose) for example, bifidobacteria were significantly increased in stool after ingestion of 5g twice a day for 6 weeks [111]. This could be the equivalent of consuming <15 g lactose (content of 1 cup of milk where 3-5 g may reach the colon in LNP subjects) twice a day [47]. The effect of smaller doses over months or years has not been studied.

Whether people with self reported LI have higher rates of gastrointestinal cancers, especially colon cancer, has not been clarified. In the study of Nicklas et al referred to above no increased rates have been observed in self reported LI persons [93]. However, we do not know the actual frequency of LNP subjects who might compensate for diminished calcium intake with small amounts of lactose ingestion. The concept may be more direct as it applies to gastric cancer which is more common in LNP populations. In this paradigm exclusion of DFs may increase the risk of cancer. In the case of gastric cancer a meta-analysis from China [112] and a another large study from Japan [113] did find reduced rates in persons (especially in men in the latter study) who consumed more DFs. These findings have partly been supported by the study of Park alluded to above [103].

Dairy Food Dependent: Increased Risk with Consumption of Lactose/Dairy Foods

A number of specific diseases will be highlighted in this section. One disease, malaria discussed above was hypothesized to derive protection from LNP status. The mechanism postulated was the formation of flavin deficient red cells which are then protected somewhat from invasion with M. Falciparum [11]. This hypothesis was not substantiated by an epidemiologic study from southern Italy. The refutation of the hypothesis rested on the parallel relationship of beta-thalassemia and glucose 6-phosphate dehydrogenase deficiency with regions located in high malaria endemicity. These hemoglobinopathies also have been postulated to protect against malaria. Alternatively, the distribution of lactase deficiency (LNP) status was similar in high and low endemicity regions [114]. In addition other studies suggested that cow's milk was actually protective against Falciparum Malaria [115].

One early link of lactose in LP persons was the predisposition to cataract formation, The pathogenesis is due to galactose absorption especially in diabetics [116]. It was suggested that the combination of high lactose intake and a presence of low galactokinase may increase the risk for cataracts by 4.1 fold [117].

Because greater quantities of milk drinking and DFs consumption occur more in countries with large LP population proportions, diseases in these societies may be linked. Hypercholesterolemia and cardiovascular diseases are one of the most common illnesses in "Western" societies. A number of studies have shown an increased rate of CVD and mortality to intakes of high- fat dairy products [118-120]. But a recent review of observational studies have failed to convincingly demonstrate an increased risk of cardiovascular disease or stroke regardless of fat content. The argument is that while low density lipoprotein is increased, so is high density lipoprotein which may counteract the increased risk. [121,122]

In evaluating relationships between dairy food consumption and a number of cancers discussed above Shrier et al found that ovarian, prostate and breast cancers had higher rates in countries with high dairy food consumption and high proportion of LP populations. At the time of publication of this observation, then available meta-analyses of the effects of dairy consumption on prostate, ovary and breast cancers appeared to have supported, the epidemiological findings, albeit not robustly [74]. The outcomes suggested that DFs consumption increased risks for both ovary and prostate but not breast cancer.

For prostate cancer the mechanism postulated is related to inhibition of conversion of vitamin D 25(OH) to 1, 25 $(OH)^2$ by calcium [123]. This in turn fails to inhibit cell proliferation. Since then, results of ongoing studies have still failed to settle conclusively whether dairy foods enhance risks or not. A recent US study reported that whole milk consumption did increase risk of the cancer [124]. Interestingly a European study examining a possible relationship of this cancer to the dominant genotype for lactase (LP status) did not find an association [125]. However, a study from India noted that lactose intolerance occurred less frequently in prostate cancer patients than controls, implying higher rates of LP [126]. An other study suggested that intake of milk, only in adolescence increased risk for advanced later life prostate cancer [127]. Following diagnosis of prostate cancer consumption of whole milk increased risk by 2.15 fold but no other DFs increased risk of metastases [128].

In the case of ovarian cancer, Cramer initially noted a marked decrease of cancer rates which correlated inversely with LNP status. He hypothesized that galactose was toxic to ovaries leading to malignant change [72]. Several reported meta-analyses did support a role for lactose and hence galactose in ovarian cancer [129-131] and a more recent Danish study did find a somewhat higher risk [132]. However, other studies failed to find any effect of dairy products on ovarian cancer [133, 134] Therefore, DF consumption still remains a disputed pathogenic factor for ovarian cancer.

The putative use of additives like estrogen in dairy foods or galactose itself have also been reported to be pathogenic in testicular cancer [135,136]. The observation that LP persons consume more DFs in general and the fact that less galactose is available for host utilization may reduce risks for ovarian and testicular cancers.

In breast cancer initial studies suggested that risk may be increased by dairy foods [137], but as outlined above meta-analyses up to the end of the first decade of this millennium failed to substantiate an increased risk [138]. A more recent meta – analysis suggested a modest protective effect of total dairy foods but not milk [139]. The included studies however, proved to be heterogenous, and as such there is still no clear proof on the role of DFs in breast cancer.

Although smoking is the most important variable affecting lung cancer several studies suggested that DFs have a deleterious effect perhaps due to the fat content on lung [74,140]. There are no clear meta-analyses.

Dairy Food Independent: Mechanisms of Effect Unclear, Evolutionary Co-marker

In the section on genetics of lactase it was pointed out that cis elements controlling target genes facilitate adaptation to environment by their amenability to soft environmental sweeps. In the review by Jones and Swallow at least 12 other genes are listed to be controlled by cis regulators [19]. Of those listed at least three interleukins (IL – 6, IL – 10, and IL – 13) are associated with immunity. They are involved with cell signaling in the inflammatory response.

In most of the diseases described the exact mechanisms are not clearly delineated but are hypothesized. Subsequent studies then either support or refute these hypotheses. Such is the case with breast, prostate and ovaries. The reasons for the vacillator outcomes may be methodological differences ar introduction of subtle biases. Nevertheless, epidemiological correlations point to differences between intake and disease and reductions with increasing LNP status.

Among disease risks apparently modified by latitude/UVB/vitamin D [68,141,142] and LP/LNP distributions [74, 143] IBDs figure prominently. In fact, there are numerous observational studies and a few interventional studies which support the benefit of vitamin D in reducing risks and improving outcome of IBD, (especially CD) [100, 144– 147]. The specific etiology of IBD is still unknown but environmental factor(s) and an abnormal immune response to microbial flora are thought to be key in pathogenesis [80]. Importantly IBD is the sine-qua-non of a complex disease in which a multitude of genetic interactive predisposing factors contribute [80]. As outlined above the first gene predisposing to CD was described at genetic locus on chromosome 16 [79]. This locus controls the interactive NOD2/CARD15 system in concert with other gene loci for Toll 2,4, 9 and autophagy sites to protect the host from luminal bacterial invasion [80]. Among pathogens which can be ingested through unpasteurized milk Brucella, Mycobacterium Tuberculosis and Listeria Monocytogenes are the most common [82].

A disease resembling CD exists in cattle, camels and goats, Johne's disease is caused by Mycobacterium Avium Paratuberculosis (MAP) [148]. This MAP organism possibly passed to humans through milk (even when pasteurized) has been sought as an etiological agent in CD and still debated [149]. The NOD2 system may have evolved in parallel to herding and milking practices and mutations described above are strongly correlated with CD [72]. However, the relationship appears to be limited largely to Caucasians, at least for the NOD2 mutations [150-152]. Most recently the CD predisposing mutation of ATG16L1 also appears to be restricted to Caucasians compared with aboriginal first nations in Canada [152]. These evolutionary changes are one potential mechanism linking IBD with lactase distributions.

This specific hypothesis of evolutionary relationship of IBD to milk drinking raises the question whether the dominant LP status itself increases IBD risk, The correlation of Crohn's disease is particularly close in geographic distributions with LP population distributions [153]. However, the answer to this question is unknown. There are only four studies which examined IBD relationships to LP/LNP status and report conflicting results. A study from Germany could not establish a protective link of CD with LNP genotype [154]. Two studies, one from north Spain [155] and one from New Zealand [156] did find that the dominant TT genotype was more represented in CD patients. However, another study from Spain failed to confirm the relationship of CD with TT genotype but did find a lower frequency in UC [157].

Evolutionary process could also be independent of either lactase genotype. An examination of genetic links between IBD and 23 other autoimmune diseases showed common predisposing genotypes [158]. These links and other yet unidentified diseases support possible associations of genes hypothesized to have evolved along with lactase.

Conclusion

Overall lactose digestion remains a great interest from three perspectives. First the symptoms related to lactose ingestion continue to occupy research efforts. With the discovery of the enzyme LPH and symptoms observed in LM subjects it was hoped that a biologically plausible explanation for irritable bowel syndrome was at hand. However, with reports that LNP persons were not the ones predominantly representing IBS cases, the LM hypothesis of IBS quickly dissipated. In the last year the geographic pattern of irritable bowel syndrome has been published and is very different from those diseases related to LP/LNP distributions [159]. In the last decade LI and LM were clearly separated and in fact the emphasis shifted from IBS symptoms to the medical hazards of DFs exclusion. Although still not clear on the best management of LI it has been shown that relatively low doses, especially consumed with food, even twice a day is readily acceptable to LM persons. Those with lactose sensitivity or LI without LM may need further assessment for other food component intolerances.

Second, the diseases and conditions associated with dairy food consumption display variable outcomes. In some cases risk of disease is increased and in others it is decreased while in still others a clear effect is difficult to prove. While the multitude of compounds found in milk naturally or added in the production (for cattle benefit) surely play some etiological roles, there appears to be less emphasis in research toward the role of lactose.

However, the lactase population dichotomy also plays a role. On a population level LNP consume less dairy foods. Any disease putatively caused by dairy would be less often observed in LNP populations. If LNP populations consume enough daily lactose which ends up being metabolized by bacteria, over time regular consumption could lead to altered and more beneficial "probiotic" type bacteria.

Finally genetic changes which in ancient times led to the emergence of LP dominance likely were accompanied by changes to protect early man against infections passed through consumption of unpasteurized milk. As well soft evolutionary sweeps which may have been responsible for LP emergence may have had effects of other similarly controlled polymorphisms.

This lactose digestive divide offers interesting teleogical (the why so) questions. In the early days of lactase/digestion of lactose, research the question was why would the world's population be divided into those who can and those who can't digest lactose? It seems this question was largely answered: because those of our ancestors who began herding cattle, camels or goats and started to milk and consume it, developed dominant genes to digest lactose. As a result it is thought that they derived increased nutritional benefits. In current times a new and important teliological question arises. Why is it that a gene for "better" nutrition which evolved 7.5 x 10^3 -10^4 years ago affects geographic distributions of so many current diseases; most having reached alarming incidence rates in the last 50-100 years. An answer may lie in altered nutrition and modernization. The evolution of protective genes against infections so that man could improve nutrition in times of scarcity may be counterproductive or altered in response in times of plenty [160].

References

[1] Gilat, T; Russo, S; Gelman-Malachi, E; Aldor, TA. Lactase in man: a nonadaptable enzyme. *Gastroenterology.* 1972 62, 1125–1127.

[2] Bersaglieri, T; Sabeti, PC; Patterson, N; Vanderploeg, T; Schaffner, SF; Drake, JA;et al. Genetic signatures of strong recent positive selection at the lactase gene. *Am J Hum Genet.* 2004 74, 1111 – 1120.

[3] Tishkoff, SA; Reed, FA; Ranciaro, A; Voight, BF; Babbitt, CC; Silverman, JS; et al Convergent adaptation of human lactase persistence in Africa and Europe. *Nat Genet.* 2007 39, 31-40.

[4] Malyarchuk, B; Derenko, M; Denisova, G; Maksimov, A; Wozniak, M; Grybowski, T; et al. Ancient links between Siberians and native Americans revealed by subtyping the Y chromosome haplogroup Q1A. *J Hum Genet.* 2011 56, 583 – 588.

[5] Forster, P; Matsumura, S. Did early humans go north or south. *Science.* 2005 308, 965 – 966.

[6] Burger, J; Kirchner, M; Bramanti, B; Haak, W; Thomas, MG. Absence of the lactase-persistence-associated allele in early Neolithic Europeans. *Proc Natl Acad Sci USA.* 2007 104, 3736-3741.

[7] Simoons, FJ. Primary adult lactose intolerance and the milking habit: aproblem in biologic and cultural interrelations. II: A culture historical hypothesis. *Am J Dig Dis.* 1970 15, 695–710.

[8] Simoons, FJ. The geographic hypothesis and lactose malabsorption: a weighing of the evidence. *Am J Dig Dis.* 1978 23, 963–980.

[9] Holden, C; Mace, R. Phylogenetic analysis of the evolution of lactose digestion in adults. *Hum Biol.* 1997 69. 605-628.

[10] Flatz, G; Rotthauwe,HW: Lactose nutrition and natural selection. *Lancet.* 1973 2, 76–77.

[11] Anderson, B; Vullo, C: Did malaria select for primary adult lactase deficiency? *Gut.* 1994 35, 1487–1489.

[12] Itan, Y; Powell, A; Beaumont, MA; Burger, J; Thomas, MG. The origins of lactase persistence in Europe. *PloS Comput Biol.* 2009 5, e1000491.doi:10.1371 /journal.pcbi.1000491.

[13] Gerbault, P; Moret, C; Currat, M; Sanchez-Mazas, A. Impact of selection and demography on the diffusion of lactase persistence. *PloS ONE,* 2009 4, e6369doi:10.1371/journal.pone.0006369.

[14] Heyer, E ; Brazier, L ; Segurel, L ; Hegay, T ; Austerlitz, F ; Quintana-Murci, L ; et al. Lactase persistence in central Asia : Phenotype, genotype, and evolution. *Hum Biol.* 2011 83, 379 – 392.

[15] Gallego Romero, I; Basu Mallick, C; Liebert, A; Crivellaro, F; Chaubey, G; Itan, Y, et al. Herders of Indian and European cattle share their predominant allele for lactase persistence. *Mol Biol Evol.* 2012 29, 249-60.

[16] Xu, L; Sun, H ; Zhang, X, ; Wang, J; Sun, D; Chen, F; et al. The G-22018 A allele matches the lactase persistence phenotype in northern Chinese populations. *Scand J Gastroenterol.* 2010 45, 168–74.

[17] Ingram, CJE; Elamin, MF; Mulcare, CA; Weale, ME Tarekegn, A; Raga, TO; et al. A novel polymorphism associated with lactose tolerance in Africa: multiple causes for lactase persistence? *Hum Genet.* 2007 120, 779 - 788.

[18] Kruse,TA; Bolund, L; Grzeschik, KH; Ropers, HH; Sjöström, H; Norén, O; et al. The human lactase-phlorizin gene is located on chromosome 2. *FEBS Lett.* 1988 240, 123-126.

[19] Jones, BL; Swallow, DM. The impact of cis-acting polymorphisms on the human phenotype. *HUGO J.* 2011 5, 13 – 23.

[20] Swallow, DM. Genetics of lactase persistence and lactose intolerance. *Ann Hum Genet.* 2003 37, 197-219.

[21] Enattah, NS ; Sahi, T ; Savilahti, E ; Terwilliger, JD; Peltonen, L; Järvelä, I. Identification of a variant associated with adult-type hypolactasia. *Nat Genet.* 2002 30, 233-237

[22] Enattah, NS; Kuokkanen, M; Forsblom, C; Natah S, Oksanen A, Jarvela I, et al. Correlation of intestinal disaccharidase activities with the C/T-13910 variant and age. *World J Gastroenterol.* 2007 13, 3508-3512.

[23] Mulcare, AC; Weale, ME; Jones, AL; Connell, B; Zeitlyn, D; Tarekegn, A; et al. The T allele of a single-nucleotide polymorphism 13.9 kb upstream of the lactase gene (LCT) (C-13.9kbT) does not predict or cause the lactase-persistence phenotype in Africans. *Am J Hum Genet.* 2004 74, 1102-1110.

[24] Imtiaz, F; Savilahti, E; Sarnesto, A; Trabzuni, D; Al-Kahtani, K; Kagevi, I; et al.The T/G 13915 variant upstream of the lactase gene (LCT) is the founder allele of lactase persistence in an urban Saudi population. *J Med Genet.* 2007 44, 4e89.

[25] Itan, Y; Jones, BL; Ingram, CJE; Swallow, DM; Thomas, MG. A worldwide correlation of lactase persistence phenotype and genotypes. *BMC Evolutionary Biology* 2010 10, 36 http://www.biomedcentral.com/1471-2148/10/36.

[26] Saviahti, A; Launiala, K; Kuitunen, P. Congenital lactase deficiency. A clinical study on 16 patients. *Arch Dis Child.* 1983 58, 246-252.

[27] Uchida, N; Sakamoto, O; Irie, M; Abukawa, D; Takeyama, J; Kure, S; et al. Two novel mutations in the lactase gene in a Japanese infant with congenital lactase deficiency. *Tohoku J Exp Med.* 2012 227, 69 – 72.

[28] Matthews, SB; Waud, JP; Roberts, AG; Campbell, AK..Systemic lactose intolerance: a new perspective on an old problem. *Postgrad Med J.* 2005 81, 167–73.

[29] Grundmann, SA ; Stratmann , E ; Brehler, R ; Luger, TA ; Ständer, S. Lactase deficiency: a potential novel aetiological factor in chronic pruritus of unknown origin. *Acta Derm Venereol.* 2011 91, 698-703.

[30] Keith, JN; Chhatrala, R. Lactose intolerance and the consumption of dairy foods. In Food and Nutritional components in focus No.3, Dietary Sugars: Chemistry, Analysis, Function and Effects Preedy, VR; ed. The *Royal Society of Chemistry,* London; 2012: 842 – 857.

[31] He, T; Venema, K; Priebe, MG; Welling, GW; Brummer, RJ; Vonk, RJ. The colonic metabolism in lactose intolerance. *Eur J Clin Invest.* 2008 38, 541-547.

[32] Duncan, SH; Louis, P; Flint, H.J.. Lactate-utilizing bacteria, isolated from human feces, that produce butyrate as a major fermentation product. *Applied and Environmental Microbiology.* 2004 70, 5810-5817.

[33] Jiang, T; Savaiano, DA. In vitro lactose fermentation by human colonic bacteria is modified by Lactobacillus acidophilus supplementation, *J Nutr.* 1997 1237, 1489 - 1495.

[34] Ito, M; Kimura, M. Influence of lactose on faecal microflora in lactose maldigesters. *Microbial Ecology in Health and Disease.* 1993 6, 73-76.

[35] Jiang, T; Savaiano, DA. Modification of colonic fermentation by bifidobacteria and pH in vitro. Impact on lactose metabolism, short-chain fatty acid, and lactate production, *Dig Dis Sci* 1997 42, 2370 - 2377.

[36] Makivuokko, HA; Saarinen, MT; Ouwehand, AC; Rautonen, NE. Effects of lactose on colon microbial community structure and function in a four-stage semi-continuous culture system, *Biosci Biotechno Biochem.* 2006 70, 2056 – 2063.

[37] Szilagyi, A; Shrier, I; Heilpern, D; Je, JS; Park, SH; Chong, G; et al. Differential impact of lactose/lactase phenotype on colonic microflora. *Can J Gastroenterol.* 2010 24, 373-379. 2010.

[38] Falony, G ; Calmeyn, T ; Leroy, F ; De Vuyst, L. fructans. Appl Environ Microbiol. 2009 75, 2312-2319.

[39] Hertzler, SR; Savaiano, DA: Colonic adaptation to daily lactose feeding in lactose maldigesters reduces lactose intolerance. *Am J Clin Nutr* 1996 64, 232–236. 1996.

[40] Tomba, C; Baldassarri, A; Coletta, M; Cesana, BM; Basilisco, G. Is the subjective perception of lactose intolerance influenced by the psychological profile? *Aliment Pharmacol Therap.* 2012 36, 660-669.

[41] Casellas, F; Aparici, A; Casaus, M; Rodríguez, P; Malagelada, JR.Subjective perception of lactose intolerance does not always indicate lactose malabsorption. *Clin Gastroenterol Hepatol.* 2010 8, 581-6.

[42] Szilagyi, A; Malolepszy, P; Yesovitch, S; Nathwani, U; Vinokuroff, C; Cohen, A; Xue, X. Inverse dose effect of pretest dietary lactose intake on breath hydrogen results and symptoms in lactase non persistent subjects. *Dig Dis Sci.* 2005 50, 2178-2182.

[43] Hertzler, SR; Huynh, BC; Savaiano, DA. How much lactose is low lactose. *J Am Diet Assoc.* 1996 96, 243-246.

[44] Oku, T; Nakamura, S; Ichinose, M. Maximum permissive dose of lactose and lactitol for transitory diarrhea, and utilizable capacity for lactose in Japanese female adults. *J Nutr Sci Vitaminol.* 2005, 51, 51-57.

[45] Montalto, M; Gallo, L; Santoro, L; D'Onofrio, F; Curigliano, V; Covino, M; et al. Low-dose lactose in drugs neither increases breath hydrogen excretion nor causes gastrointestinal symptoms. *Aliment Pharmacol Therap.* 2008 28, 1003 – 1012.

[46] Suarez, FL; Savaiano, DA; Levitt, MD. A comparison after the consumption of milk or lactose-hydrolyzed milk by people with self-reported severe lactose intolerance, *N Engl J Med.* 1995 333,1 - 4.

[47] Suarez, FL; Savaiano, D ; Arbisi, P ; Levitt, MD. Tolerance to the daily ingestion of two cups of milk by individuals claiming lactose intolerance, *Am. J. Clin. Nutr.* 1997 65, 1502 – 1506.

[48] Savaiano, DA; Boushey, CJ; McCabe, GP. Lactose intolerance symptoms assessed by meta-analysis: A grain of truth that leads to exaggeration. *J Nutr.* 2006 136, 1107 - 1113.

[49] Vesa, TH; Seppo,LM; Marteau, PR; Sahi, T; Korpela, R. Role of irritable bowel syndrome in subjective lactose intolerance *Am J Clin Nutr.* 1998 67, 710 – 715.

[50] DiStefano, M; Terulla, V; Tana, P; Mazzocchi, S; Romero, E; Corazza, GR. Genetic test for lactase non-persitence and hydrogen breath test: Is genotype better than phenotype to diagnose lactose malabsorption? *Dig Liv Dis.* 2009 41, 474-479.

[51] Maiuri, L; Raia, V; Potter, J; Swallow, D; Ho, MW; Fiocca, R; et al. Mosaic pattern of lactase expression by villous enterocytes in human adult-type hypolactasia. *Gastroenterology* 1991; 100: 359-369.

[52] Furnari, M ; Bonfanti, D ; Parodi, A ; Franze, J ; Savarino, E ; Bruzzone, L ; et al. Comparison Between Lactose Breath Test and Quick Test on Duodenal Biopsies for Diagnosing Lactase Deficiency in Patients With Self-reported Lactose Intolerance. *J Clin Gastroenterol.* 2013 47,148-152.

[53] Newcomer, AD; McGill, DB; Thomas, PJ; Hofmann, AF. Prospective comparison of indirect methods for detecting lactase deficiency. *N Engl J Med.* 1975 293, 1232-1236.

[54] Arola, H. Diagnosis of hypolactasia and lactose malabsorption. *Scand J Gastroeneterol.* 1994 202, 26-35.

[55] Metz, G; Jenkins, DJA; Peters, TJ; Newman, A; Blendis, LM. Breath hydrogen as a diagnostic method for hypolactasia. *Lancet.* 1975 I, 1155-1157.

[56] Szilagyi, A. *Lactase Phenotype and Relationships to Diseases.* In Genetics Research, Volume 6, Nova Science Publishers Inc, New York, Editor; Kevin V Urbano. 2011, 287 – 308.

[57] Strocchi, A; Corazza, G,; Ellis, CJ; Gasbarrini, G; Levitt, MD. Detection of malabsorption of low doses of carbohydrate: accuracy of various breath H2 criteria. *Gastroenterology.* 1993 105,1404–1410.

[58] Waud, JP; Matthews, SB; Campbell, AK. Measurement of breath hydrogen and methane, together with lactase genotype, defines the current best practice for investigation of lactose sensitivity. *Ann Clin Biochem.* 2008 45(Pt 1), 50-58.

[59] Romagnuolo, J; Schiller, D; Bailey, RJ. Using breath tests wisely in a gastroenterology practice: An evidence-based review of indications and pitfalls in interpretation. *Am J Gastroeneterol.* 2002 97, 1113-1126.

[60] Marton, A; Xue, X; Szilagyi, A. Meta-analysis: The Diagnostic Accuracy of Lactose Breath Hydrogen or Lactose Tolerance Tests for Predicting the North European lactase polymorphism C/T – 13910.*Alimentary Pharmacol Therap.* 2012 35, 429-440.

[61] Moan, J; Porojnicu, AC; Dahlback, A: Ultraviolet radiation and malignant melanoma. *Adv Exp Med Biol.* 2008 624, 104–416.

[62] Garg, M; Lubel, JS; Sparrow, MP; Holt, SG; Gibson, PR. Review article: vitamin D and inflammatory bowel disease – established concepts and future directions. *Aliment Pharmacol Therap.* 2012 36, 324 -344.

[63] Zitterman, A. Vitamin D in preventive medicine: are we ignoring the evidence ? *Br J Nutr.* 2003 89, 552-572.

[64] Cantorna, MT; Mahon, BD. Mounting evidence for vitamin D as an environmental factor affecting autoimmune disease prevalence. Exp Biol Med (Maywood). 2004 229, 1136–1142.

[65] Grant, WB. A meta-analysis of second cancers after a diagnosis of non-melanoma skin cancer: Additional evidence that solar ultraviolet –B irradiance reduces the risk of internal cancers. *J Steroid Biochem Mol Biol.* 2007 103, 668-674.

[66] Mohr, SB; Garland, CF; Gorham, ED; Grant, WB; Garland, FC. Ultraviolet B and incidence rates of leukemia worldwide. *Am J Prev Med.* 2011 41, 68 – 74.

[67] Simpson, S Jr; Blizzard, L; Otahal, P; Van der Mei, I; Taylor, B.Latitude is significantly associated with the prevalence of multiple sclerosis: a meta-analysis. *J Neurol Neurosurg Psychiatry.* 2011 82, 1132-1141.

[68] Peyrin-Biroulet, I; Oussalah, A; Bigard, M-A. Crohn's disease: the hot hypothesis. *Med Hypotheses.* 2009 73, 94–96.

[69] Szilagyi, A; Leighton, H; Burstein, B; Shrier, I. Significant positive correlation between sunshine and lactase nonpersistence in Europe may implicate both in similarly altering risks for some diseases. *Nutr Cancer.* 2011 63, 991 – 999. 2011

[70] Szilagyi, A. Determinants of Geographic Patterns of Diseases: Interaction of Lactose/Lactase status and Sunshine Exposure. *Med Hypotheses.* 2010 75, 466-470.

[71] International Agency for Research Group. Dietary fiber,transit – time, faecal bacteria, steroids, and colon cancer in two Scandinavian populations. *Lancet.* 1977 2, 207 – 211.

[72] Cramer, DW. Lactase persistence and milk consumption as determinants of ovarian cancer risk. *Am J Epidemiol.* 1989 130, 904 -910.

[73] Nanji, AA; Denardi, FG. Primary adult lactose intolerance protects against development of inflammatory bowel disease. *Med Hypotheses.* 1986. 19, 1-6.

[74] Shrier, I; Szilagyi, A; Correa, JA. Impact of lactose containing foods and the genetics of lactase on diseases: An analytical review of population data. *Nutr Cancer.* 2008 60, 292-300.

[75] Alfonso-Sanchez, MA ; Perez-Miranda, AM ; Garcia-Obregon, S ; Pena, JA. An evolutionary approach to the high frequency of the delta F508 CFTR mutation in European populations. *Med Hypotheses.* 2010 74, 989 – 992.

[76] Modiano, G; Ciminellui, BM; Pignatti, PF. Cystic fibrosis and lactase persistence : a possible correlation. *Eur J Hum Genet.* 2007 15, 255-259.

[77] Madry, E; Fidler, E; Sobczynska-Tomaszewska, A; Lisowska, A; Kryzanowska, P; Pogorozelski, A; et al. Mild CFTR mutations and genetic predisposition to lactase persistence in cystic fibrosis. *Eur J Hum Genet.* 2011 19, 748 -752.

[78] Truedsson, M; Carlson, J; Simren, M; Ohlsson, B. Polymorphisms in the oxytocin promoter region in patients with lactase non-persistence is not related to symptoms. *BMC Gastroenterol.* 2009 9, 90.

[79] Hugot, JP ; Chamaillard, M ; Zouali, H ; Lesage, S; Cézard, JP; Belaiche, J; et al. Association of NOD2 leucine-rich repeat variants with susceptibility to Crohn's disease. *Nature.* 2001 41, 599–603.

[80] Abraham, C; Cho, JH. Inflammatory bowel disease. *N Engl J Med.* 2009 361, 2066-2078.

[81] Abraham, C; Medzhitov, R. Interactions between the host innate immune system and microbes in inflammatory bowel disease. *Gastroenterology.* 2011 140, 1729 -1737.

[82] Gasche C, Nemeth M, Grundtner P, Willheim – Polli C, Ferenci P, Schwarzenbacher R. Evolution of Crohn's disease-associated Nod2 mutations. *Immunogenetics.* 2008 60, 115 – 120.

[83] Sacerdote, C; Guarrera, S; Smith, GD; Grioni, S; Krogh, V; Masala, G; et al. Lactase persistence and bitter taste response: Instrumental variables and Mendelian randomization in epidemiologic studies of dietary factors and cancer risk, *Am J Epidemiol.* 2007 166, 576 – 581.

[84] Norat, T; Riboli, E. Dairy products and colorectal cancer: a review of possible mechanisms and epidemiological evidence. *Eur J Clin Nutr.* 2003 57, 1–17.

[85] Yuen, AW; Jablonski, NG: Vitamin D. in the evolution of human skincolor. *Med Hypotheses.* 2010 74, 39–44.

[86] Gleiberman, L. Sodium, blood pressure and ethnicity: what have we learned? *Am J Hum Biol.* 2009 21, 679 – 686.

[87] Weder, AB. Genetics and hypertension. *J Clin Hypertension.* 2007 9, 217 – 223.

[88] Suchy, FJ; Brannon, PM; Carpenter, TO; Fernandez, JR; Gilsanz, V; Gould, JB; et al. National Institutes of Health Consensus Development Conference: Lactose intolerance and health. *Ann Int Med.* 2010 152, 792-796.

[89] Brannon PM, Carpenter TO, Fernandez JR, Gilsanz V, Gould JB; Hall KE, et al. NIH Consensus Development Conference Statement: Lactose Intolerance and Health. *NIH Consens State Sci Statements.* 2010 27, 1-27.

[90] Wilt TJ, Shaukat A, Shamliyan T, Taylor BC, MacDonald R, Tacklind J, et al. I, Lactose intolerance and health.*Evid Rep Technol Assess.* 2010 192,1-410.

[91] Koek WNH, van Meurs JB, van der Eerden BCJ, Rivadeneira F, Zillikens MC, Hofman A, et al. The T-13910C polymorphism in the lactase phlorizin hydrolase gene is associated with differences in serum calcium levels and calcium intake. *J Bone Min Res,* 2010 25, 1980-1987.

[92] Louie JC, Flood VM, Rangan AM, Burlitsky G, Gill TP, Gopinath B, et al. Higher regular fat dairy consumption is associated with lower incidence of metabolic syndrome but not type 2 diabetes. *Nutr Metab Cardiovasc Dis.* 2012; doi: 10.1016/j.numecd.2012.08.004.

[93] Nicklas TA, Haiyan, Q; Hughes SO, He M, Wagner SE, Foushee HR, et al. Self – percieved lactose intolerance results in lower intakes of calcium and dairy foods and is associated with hypertension and diabetes in adults. *Am J Clin Nutr.* 2011 94, 191-198.

[94] Chen M, Pan A, Malik VS, Hu FB. Effects of dairy intake on body weight and fat: meta-analysis of randomized controlled trials. *Am J Clin Nutr.* 2012 96, 735-747.

[95] Kettunen J, Silander K, Saarela O, Amin N, Müller M, Timpson N, et al. European lactase persistence genotype shows evidence of association with increase in body mass index. *Hum Mol Genet.* 2010 19,1129-1136.
[96] Grant, WB. A meta-analysis of second cancers after a diagnosis of non-melanoma skin cancer: Additional evidence that solar ultraviolet –B irradiance reduces the risk of internal cancers. *J Steroid Biochem Mol Biol.* 2007; 103: 668-674.
[97] Grant, WB: How strong is the evidence that solar ultraviolet B and vitamin D reduce the risk of cancer? *Dermato-Endocrinology.* 2009 1, 17–24.
[98] van Leeuwen, MT; Turner, JJ; Falster, MO; Meagher, NS; Joske, DJ; Grulich, AE; et al. Latitude gradients for lymphoid neoplasm subtypes in Australia support an association with ultraviolet radiation exposure. *Int J Cancer.* 2013, Feb 5. doi: 10.1002/ijc.28081. [Epub ahead of print]
[99] Mohr, SB; Garland, CF; Gorham, ED; Grant, WB; Garland, FC. Ultraviolet light B and incidence rates of leukemia world wide. *Am J Prev Med.* 2011 41, 68 – 74.
[100] Ananathakrishnan, AN; Khalili H; Higuchi, LM; Bao, Y; Korzenik, JR; Giovannucci, EL; et al. Higher predicted vitamin D status is associated with reduced risk of Crohn's disease. *Gastroenterology.* 2012 142, 482 - 489.
[101] Simpson, S Jr; Blizzard, L; Otahal, P; Van der Mei, I; Taylor, B. Latitude is significantly associated with the prevalence of multiple sclerosis: a meta-analysis. *J Neurol Nerosurg Psychiatry.* 2011 82, 1132-1141.
[102] Keith, JN; Chhatrala, R. Lactose Intolerance and the consumption of dairy foods in Food and Nutritional Components in Focus No. 3, Dietary Sugars: Chemistry, Analysis, Function and Effects. Preedy VR ed. *The Royal Society of Chemistry,* London. 2012 Ch 48, 842-857.
[103] Park, Y; Leitzmann, MF; Subar, Bao, Y, Korzenik, JR, Giovannucci, EL; et al. *Dairy Food, Calcium, and Risk of Cancer in the NIH-AARP Diet and Health Study Arch Intern Med.* 2009 169, 391-401.
[104] Zhang, X; Giovannucci, E. Calcium, vitamin D and colorectal cancer chemoprevention. *Best Pract Res Clin Gastroenterol.* 2011 25, 485-494.
[105] Wallace, K; Baron, JA; Cole, BF; Sandler, RS; Karagas, MR; Beach, MA; et al. Effect of calcium supplementation on the risk of large bowel polyps. *J Natl Cancer Inst.* 2004 96, 921-925.
[106] Grau, MV; Baron, JA; Sandler, RS; Wallace, K; Haile, RW; Church, TR; et al. Prolonged effect of calcium supplementation on risk of

colorectal adenomas in a randomized trial. *J Natl Cancer Inst.* 2007 99, 129-136.

[107] Grau, MV; Baron, JA; Sandler, RS; Haile, RW; Beach, ML; Church, TR; et al. Vitamin D, calcium supplementation, and colorectal adenomas: results of a randomized trial. *J Natl Cancer Inst.* 2003 95, 1765-1771.

[108] Szilagyi, A; Nathwani, U; Vinokuroff, C; Correa, JA; Shier, I. Evaluation of relationships among national colorectal cancer mortality rates, genetic lactase non persistence status and per capita yearly milk and milk product consumption. *Nut Cancer.* 2006 55,151-156.

[109] Szilagyi, A. Lactose: A potential prebiotic. *Aliment Pharmacol Therap.* 2002 16,1591-1602.

[110] Commane, D; Hughes, R; Short, C; Rowland, I. The potential mechanisms involved in the anti-carcinogenic action of probiotics. *Mutat Res.* 2005 591, 276-289.

[111] Bouhnik, Y; Attar, A; Joly, FA; Riottot, M; Dyard, F; Flourié, B. Lactulose ingestion increases faecal bifidobacterial counts: a randomized double-blind study in healthy humans. *Eur J Clin Nutr.* 2004 58, 462-466

[112] Huang, YX ; Qin, LQ ; Wang, PY. *Meta-analysis of the relationship between dairy product consumption and gastric cancer.* Zhonghua Yu Fang Yi Xue Za Zhi. 2009 43, 193-196. (article in Chinese with English Abstract).

[113] Pham, TM; Fujino, Y; Kikuchi, S; Tamakoshi, A; Matsuda, S; Yoshimura, T. Dietary patterns and risk of stomach cancer mortality: the Japan collaborative cohort study. *Ann Epidemiol.* 2010 20, 356-363.

[114] Meloni, T ; Colombo, C ; Ruggiu, G ; Dessena, M ; Meloni, GF. Primary lactase deficiency and past malarial endemicity in Sardinia. *Ital J Gastroenterol Hepatol.* 1998 30, 490-493.

[115] Lokki, AI; Järvelä, I; Israelsson, E; Maiga, B; Troye-Blomberg, M; Dolo, A; et al. Lactase persistence genotypes and malaria susceptibility in Fulani of Mali. *Malar J.* 2011 10, 9.

[116] Birlouez-Aragon, I; Ravelontseheno, L; Villate-Cathelineau, B; Cathelineau, G Abitbol, G. Disturbed galactose metabolism in elderly and diabetic humans is associated with cataract formation. *J Nutr.* 1993 123, 1370-1376.

[117] Jacques, PF; Phillips, J; Hartz, SC; Chylack, LT. Lactose intake, galactose metabolism and senile cataracts. *Nutr Res.* 1990 10, 255 -265.

[118] Seely, S. Possible connection between milk and coronary heart disease: the calcium hypothesis. *Med Hypotheses.* 2000 54, 701-3.

[119] Nettleton, JA; Steffen, LM; Loehr, LR; Rosamond, WD; Folsom, AR. Incident heart failure is associated with lower whole-grain intake and greater high-fat dairy and egg intake in the Atherosclerosis Risk in Communities (ARIC) study. *J Am Diet Assoc.* 2008 108, 1881-1887.

[120] van Aerde, MA; Soedamah-Muthu, SS; Geleijnse, JM; Snijder, MB; Nijpels, G; Stehouwer, CD; et al. Dairy intake in relation to cardiovascular disease mortality and all-cause mortality: the Hoorn Study. *Eur J Nutr.* 2013 52, 609-616.

[121] Huth, PJ; Park, KM. Influence of dairy product and milk fat consumption on cardiovascular disease risk: a review of the evidence. *Adv Nutr.*2012 3,266-85.

[122] Louie, JC; Flood, VM; Burlutsky, G; Rangan, AM; Gill, TP; Mitchell, P. Dairy consumption and the risk of 15-year cardiovascular disease mortality in a cohort of older Australians. *Nutrients.* 2013 5, 441-454.

[123] Skowronski, RJ; Peehl, PY; Feldman, D. Vitamin D and prostate cancer: 1,25 dihydroxyvitamin D3 receptors and actions in human prostate cancer cell lines. *Endocrinology.* 1993 132, 1952 – 1960.

[124] Song, Y ; Chavarro, JE ; Cao, Y ; Qiu, W ; Mucci, L ; Sesso, HD ; et al. Whole Milk Intake Is Associated with Prostate Cancer-Specific Mortality among U.S. Male Physicians. *J Nutr.* 2013 143, 189-196.

[125] Travis, RC; Appleby, PN; Siddiq, A; Allen, NE; Kaaks, R; Canzian, F;et al. Genetic variation in the lactase gene, dairy product intake and risk for prostate cancer in the European prospective investigation into cancer and nutrition. *Int J Cancer.* 2013, 132, 1901-1910.

[126] Agarwal, MM; Rana, SV; Mandal, AK; Malhotra, S; Khandelwal, N; Kumar, S; et al. Lactose intolerance in prostate cancer patients: incidence and associated factors. *Scand J Gastroeneterol* 2008;43(3):270-6

[127] Torfadottir, JE ; Steingrimsdottir, L ; Mucci, L; Aspelund, T ; Kasperzyk, JL ; Olafsson, O ; et al. Milk intake in early life and risk of advanced prostate cancer. *Am J Epidemiol.* 2012 175, 144-153.

[128] Pettersson, A; Kasperzyk, JL; Kenfield, SA; Richman, EL; Chan, JM; Willett,WC; et al. Milk and dairy consumption among men with prostate cancer and risk of metastases and prostate cancer death. *Cancer Epidemiol Biomarkers Prev.* 2012 21,428-436.

[129] Qin, LQ; Xu, JY; Wang, PY; Hashi, A; Hoshi, K; et al.: Milk/dairy products consumption, galactose metabolism and ovarian cancer: meta-analysis ofepidemiological studies. *Eur J Cancer Prev.* 2005 14, 13–19.

[130] Larsson, SC; Orsini, N; Wolk, A: Milk, milk products and lactose intake and ovarian cancer risk: a meta-analysis of epidemiological studies. *Int JCancer* 118, 431–441, 2006.

[131] Genkinger, JM; Hunter, DJ; Spiegelman, D; Anderson, KE; Arslan, A; et al.: Dairy products and ovarian cancer: a pooled analysis of 12 cohort studies. *Cancer Epidemiol Biomarkers Prev.* 2006 15, 364–372.

[132] Faber, MT; Jensen, A; Søgaard, M; Høgdall, E; Høgdall, C; Blaakaer, J; et al. Use of dairy products, lactose, and calcium and risk of ovarian cancer - results from a Danish case-control study. *Acta Oncol.* 2012 51, 454-464.

[133] Mommers, M; Schouten, LJ; Goldbohm, RA; van den Brandt, PA. Dairy consumption and ovarian cancer risk in the Netherlands Cohort Study on Diet and Cancer. *Br J Cancer.* 2006 94, 165–170.

[134] Schulz, M ; Nöthlings, U ; Allen, N ; Öland-Moret, NC ; Agnoli, C ; Engeset, D ; et al. No association of consumption of animal foods with risk of ovarian cancer. *Cancer Epidemiol Biomark Prev. 2007* 16, 852-855.

[135] Davies, TW; Palmer; CR; Lipscombe, JM. Adolescent milk, dairy product and fruit consumption and testicular cancer. *Br J Cancer.* 1998 74, 657- 660.

[136] Stang, A; Ahrens, W; Baumgrdt-Elms, C. Adolescent milk fat and galactose consumption and testicular cell cancer. *Cancer Epidemiol Biomark Prev.* 2006 15:,2189 – 2195.

[137] Ganmaa, D; Sato, A. The possible role of female sex hormones in milk from pregnant cows in the development of breast, ovarian and corpus uteri cancers. *Med Hypotheses.* 2005 65,1028–1037.

[138] Krog, V; Berrino, F; Sieri, S; Grioni, S; Tjønneland, A; et al. Meat, eggs, dairy products, and risk of breast cancer in the European Prospective Investigation into Cancer and Nutrition (EPIC) cohort. *Am J Clin Nutr.* 2009 90, 602-612.

[139] Dong, JY; Zhang, L; He, K; Qin, LQ. Dairy consumption and risk of breast cancer: a meta-analysis of prospective cohort studies. *Breast Cancer Res Treat.* 2011 127, 23-31.

[140] Sankaranarayanan, R; Varghese, C; Duffy, SW; Padmakumary, G; Day, NE Nair MK. A case-control study of diet and lung cancer in Kerala, south India. *Int J Cancer.* 1994 58, 644-649.

[141] Sonnenberg, A; McCarty, DJ; Jacobsen, SJ. Geographic variation of inflammatory bowel disease within the United States. *Gastroenterology.* 1991 100, 143-149.

[142] Shivananda, S; Lennard-Jones, J; Logan, R; Logan, R Fear, N; Price, A; Carpenter, L; et al. Incidence of inflammatory bowel disease across Europe : is there a difference between north and south ? Results of the European collaborative study on inflammatory bowel disease (EC-IBD). *Gut.* 1996 39, 690-697.

[143] Juste, RA. Crohn's disease and ruminant farming. Got lactase? *Med Hypotheses.* 2010 75, 7-13.

144) Nerich, V; Jantchou, P; Boutron-Ruault, M-C; Monnet, E; Weill, A; Vanbockstael, V; et al. Low exposure to sunlight is a risk factor for Crohn's disease. *Aliment Pharmacol Therap. 2011* 33, 940- 945.

[145] Ulitsky, A; Ananthakrishnan, AN; Naik, A; Skaros, S; Zadvornova, Y; Binion, DG; et al.Vitamin D deficiency in patients with inflammatory bowel disease: association with disease activity and quality of life. *J Parent Ent Nutr.* 2011 35, 308- 316.

[146] Pappa, HM; Grand, RJ; Gordon, CM. Report on the vitamin D status of adult and pediatric patients with inflammatory bowel disease and its significance for bone health and disease. *Inflamm Bowel Dis.* 2006 12, 1162-1174.

[147] Jorgensen, SP; Agnholt, J; Glerup, H; Lyhne, S; Villadsen, GE; Hvas, CL; et al. Clinical trial : vitamin D3 treatment in Crohn's disease – a randomized double-blind placebo- controlled study. *Aliment Pharmacol Therap.* 2010 32, 377-383.

[148] Waddell, LA; Rajić, A; Sargeant J; Harris, J; Amezcua, R; Downey, L; et al. The zoonotic potential of Mycobacterium avium spp. paratuberculosis: a systematic review. *Can J Public Health.* 2000 99, 145-155.

[149] Feller, M; Huwiler, K; Stephan, R; Altpeter, E; Shang, A; Furrer, H; et al. Mycobacterium avium subspecies paratuberculosis and Crohn's disease: a systematic review and meta-analysis. *Lancet Infect Dis.* 2007 7,607-613.

[150] Parkes, M. Evidence from genetics for a role of autophagy and innate immunity in IBD pathogenesis. *Dig Dis.* 2012 30, 330-333.

[151] Naser, SA ; Arce, M ; Khaja, A ; Fernandez, M ; Naser, N ; Elwasila, S ; et al. Role of ATG16L, NOD2 and IL23R in Crohn's disease pathogenesis. *World J Gastroenterol.* 2012 18, 412-424.

[152] Murdoch, TB; Bernstein, CN; El-Gabalawy, H; Stempak, JM; Sargent, M; Elias, B; et al. Prevalence of genetic variants associated with inflammatory bowel disease in a healthy First Nations cohort. *CMAJ.* 2012 184, E435-41.

[153] Szilagyi, A; Leighton, H; Burstein, B; Xue, X. Global parameters and geography of inflammatory bowel diseases. *Can J Gastroenterol.* 2013 27 SupplA, 138A.:

[154] Buning, C; Ockenga, JK; Krüger S, Jurga J, Baier P, Dignass A, et al. The C/C(-13910) and G/G9-22018) genotypes for adult-type hypolactasia are not associated with inflammatory bowel disease. *Scand J Gastroenterol.* 2003 38, 538-542.

[155] Elguezabal, N; Chamorro, S; Molina, E; Garrido, J; Izeta, A; Rodrigo, L; et al. Lactase persistence, NOD2 status and Mycobacterium avium subsp. Paratuberculosis infection associations to inflammatory bowel disease. *Gut Pathogens.* 2012 4, 6.

[156] Nolan, DJ; Han, DY; Lam, WJ; Morgan, AR; Fraser, AG; Tapsell, LC; et al. Genetic adult lactase persistence is associated with risk of Crohn's disease in a New Zealand population *BMC Res Notes.* 2010 3, 339.

[157] Juste, RA; Elguezabal, N; Chamorro, S; Molina, E; Garrido, JM; Riestra, S; et al. Associations between CARD15 polymorphisms, MAPDNA in blood and lactase persistence in a Crohn's disease case-control study in north Spain. Minnesota: In *Proceedings of the 10th International Colloquim on Paratuberculosis.* 2009 269, 9-14.

[158] Lees, CW; Barrett, JC; Parkes, M; Satsangi, J. New IBD genetics: common pathways with other diseases. *Gut.* 2011 60,1739-53.

[159] Lovell, RM; Ford, AC. Global prevalence of and risk factors for irritable bowel syndrome: a meta-analysis.*Clin Gastroenterol Hepatol.* 2012 10, 712-721.

[160] Di Rienzo, A; Hudson, RR. An evolutionary framework for common diseases: the ancestral – susceptibility model. *Trends Genet.* 2005 21, 596-601.

In: Lactose
Editors: David Green and Emma Lee
ISBN: 978-1-62808-198-5

Chapter 3

Whey Lactose Fermentation by Kefir Immobilized on Agro-industrial Wastes

Magdalini Soupioni
Nuclear Chemistry Laboratory, Department of Chemistry,
University of Patras, Patras, Greece

Abstract

Whey is the main liquid by-product of the dairy manufactures, in which a large quantity of the milk lactose is removed. So, whey consists of lactose (4.8-5.3%), proteins (0.8-1.6%), fats (0.3-1.3%), vitamins (0.8%) and minerals. The fat and protein components are rather easily recovered, but lactose responsible for 90% of whey's COD remains unused.

The last two decades, the natural mixed culture kefir was chosen for whey lactose exploitation, because its microbiological composition ensures simultaneous alcoholic and lactic acid whey fermentations resulting to useful products formation like ethanol, lactic and propionic acid. Also, the immobilization of kefir cells on various natural food grade supports and the preparation of biocatalysts suitable for easy and fast whey fermentation have received considerable attention by researchers.

In this chapter the whey lactose fermentation kinetics by kefir immobilized on Delignified Cellulosic Materials (DCM), Gluten Pellets (GP), Brewer's Spent Grains (BSG), Malt Spent Rootlets (MSR) and

Corn Starch Gel (CSG) was examined in relation to free kefir cells (FC) for comparative reasons. Also, lactose uptake rate by immobilized and free kefir cells during fermentation was monitored by adding ^{14}C-labelled lactose in whey and the produced ethanol was determined.

The biocatalysts prepared by immobilization of kefir cells on agro-industrial wastes were proved effective for whey fermentation at 30°C and pH 5.5, reducing significantly fermentation time and increasing lactose uptake rate in comparison to free cells. Specifically, using cheap, abundant and food grade BSG or DCM biocatalysts for whey fermentation is potentially acceptable by industries as the fermentation time was estimated only about 7 hours in that case. The highest alcohol concentration was observed by MSR biocatalyst.

Introduction

Whey is the main liquid by-product of the dairy activities, in which a large quantity of the lactose of the milk is removed. It is of great nutritious value as it consists of mainly lactose (4.8-5.3%), proteins (0.8-1.6%), fats (0.3-1.3%) vitamins (0.8%) and minerals. The world whey production is over 160 million tones per year, showing a 1–2% annual growth rate (OECD-FAO 2008; Smithers, 2008). But, in many countries is usually discarded as a waste in the environment, representing a grave pollutant, because of its high BOD_5 (40-70 g/L) and COD (60-80 g/L) content, making the biological treatment difficult.

The fat and protein components of whey are rather easily recovered, but lactose responsible for 90% of whey's COD normally remains unused. Last thirty years extensive research efforts were undertaken by many investigators for whey utilization using microorganisms that ferment lactose. The improvement of whey lactose fermentation was necessary, in order to reduce whey polluting load and to produce possibly useful bioproducts. In that way the streams of whey could be used as an abundant and renewable potential raw material for microbial fermentations (Panesar et al., 2007; Kosseva et al., 2009). Depending on the microorganisms involved, lactose contained in whey was fermented to ethanol, biogas (methane), bio-hydrogen, single cell protein (SCP), beverages, lactic acid, acetic acid, bio-polymers, bio-surfactants, etc. For examble, the fermentation of whey by *Kluyveromyces marxianus* and *Candida pseudotropicalis* has been used to produce alcohol (Soo et al., 1994; Ghaly et al., 1995; Silveira et al., 2005; Zafar et al., 2006; Kargi et al., 2006; Koutinas et al., 2007; Oznihci et al., 2007; Oznihci et al., 2008; Sansonetti S., et al., 2009). Also, the bioconversion of whey by *Kluyveromyces fragilis*

(Chaly et al., 2004; Chaly et al., 2005) has been used to produce SCP and the fermentation of whey by *Lactobacillus delbrueckii* (Arasaratnam et al., 1996), *Lactobacillus casei* (Panesar et al., 2010), *Lactobacillus* and *Tetragenococcus,* (Zaied et al., 2009), has been used to produce lactic acid. Moreover, whey fermentation by *Aspergillus niger* has been used to produce citric acid (Samragy et al., 1996).

But, almost all the before mentioned microorganisms as well as many other various species of yeasts and bacteria co-exist in a symbiotic association in the natural mixed culture of kefir (Figure 1), embedded into a matrix consistent of polysaccharides, named kefiran (Garrote et al., 1997; Pintado et al., 1996). Due to its microbiological composition, ensuring simultaneous alcoholic and lactic acid fermentations, kefir was thought to be ideal for the exploitation of whey lactose (Athanasiadis et al., 2002; Athanasiadis et al., 2001; Harta et al., 2004). So, during last two decades many researchers used fermentation of whey by kefir to produce mainly potable alcohol (Athanasiadis et al., 2002), and whey alcoholic beverages (Paraskevopoulou et al., 2003; Koutinas et al., 2005; Dragone et al., 2009; Magalhaes et al., 2010; Pescuma et al., 2010) with low cost. Simultaneously, as whey lactose fermentation by kefir was strongly advantaged by cell immobilization techniques, several such techniques and support materials have already been proposed (Athanasiadis et al., 1999; Plessas et al., 2005; Athanasiadis et al., 2004; Dimitrellou et al., 2007). Finally, natural organic materials of food grade have been prevailed as supports for kefir cells immobilization, in order to avoid undesirable residues in the final products of whey fermentation.

Also, as has been done before (Soupioni et al., 1998), in an attempt to confirm directly the positive effect of immobilized kefir on whey lactose fermentation rate, ^{14}C-labelled lactose (Figure 2) was used for monitoring lactose uptake rate by free kefir cells and by immobilized ones (Golfinopoulos et al., 2009; Golfinopoulos et al., 2011; Golfinopoulos et al., 2012; Soupioni et al., in press). There, it was found that lactose uptake rate is strongly correlated to fermentation rate and increased during whey fermentation by kefir cells immobilized on delignified cellulosic materials, (DCM), gluten pellets, (GP), brewer's spent grains, (BSG), and malt spent rootlets, (MSR), separately, in comparison with fermentation by free kefir cells (FC). Also, in all cases the highest fermentation rate and lactose uptake rate by immobilized kefir was recorded at temperature of 30°C and at 5.5 pH value, except for DCM where the best results observed at 5 pH value.

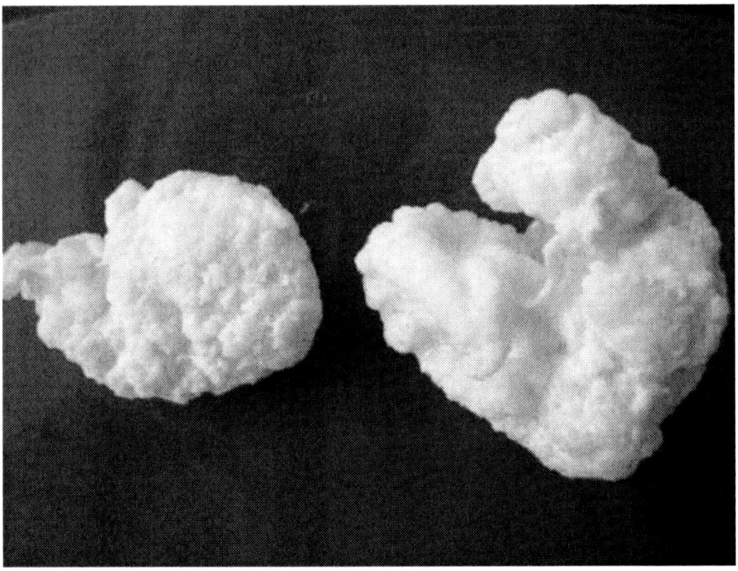

Figure 1. Kefir grains (photo by M. Soupioni).

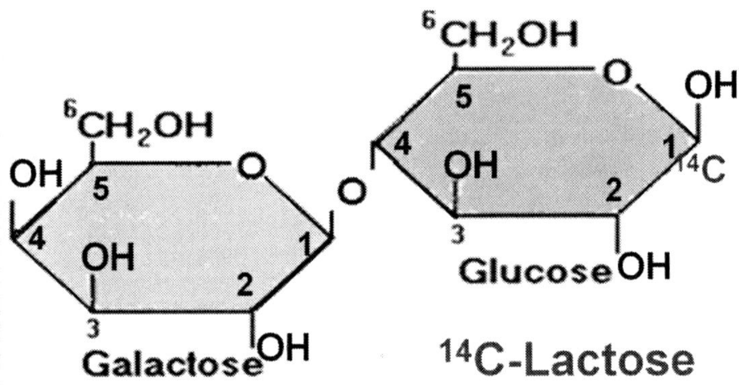

Figure 2. The ^{14}C-labelled lactose.

In the present chapter one more biocatalyst prepared by "immobilization" of kefir on it was used for whey lactose fermentation too. The Corn Starch Gel (CSG) made from "expired" packages of corn flower was used as carrier for entrapment of kefir cells. The CSG has been already used for yeast cell entrapment in order to prepare a proper biocatalyst for wine fermentation (P. Kandylis et al., 2008). The optimum conditions, like pH and temperature, for lactose uptake rate by the new biocatalyst were examined and the produced ethanol was determined during whey fermentation. The results were compared with results by DCM, GP, BSG and MSR biocatalysts as well as by FC in an attempt to propose the best system for fast, easy and low cost removal of lactose from whey.

Materials and Methods

Microorganism and Cell Growth

Kefir, a commercial product usually used to produce kefir drink, was employed in the present study. Kefir grains (Figure 1) were preserved into 1L fresh pasteurised full cream milk, which was renewed every week, at approximately 4°C. Cell growth of kefir biomass took place in a sterilized synthetic medium (2% lactose, 0.4% yeast extract, 0.1% $(NH_4)_2SO_4$, 0.1% KH_2PO_4 and 0.5% $MgSO_4.7H_2O$) at 30°C under aerobic conditions.

Materials and Media

Liquid cheese whey used for all fermentation experiments was obtained from the regional dairy industry "Agricultural Cooperative Union of Kalavryta" (Kalavryta, Greece). It remained after the production of feta cheese and contained about 5% lactose, 0.8% proteins and its pH value was 6.5. Corn flower and white wheat flour from "expired" packages used for the preparation of GP and CSG were manufactured by Hellenic Allatini Co and Giotis Co, Greece, respectively. Corn flower contained 87% starch and white wheat flour contained 13% protein. Wood sawdust from beech tree obtained from a carpenter's shop and was used for preparation of DCM. The BSG and MSR were obtained from the Athenian Brewery S.A. and used as cells supports, because they have food grade purity. The BSG contained about (% w/w on dry matter) 21.6 of crude protein, 6.8 of fats, 17 of cellulose, 22 of lignin, 28 of non-cellulosic polysaccharides and 4.7 of ash. The MSR consist of about (% w/w on dry matter) 36.2 protein contents, 12.1 cellulose, 18.4 fibres, 13.3 carbohydrates, 3.7 glucose, 1.0 polyphenols, 0.6 fats and 3.0 ashes. The compositions of BSG, MSR and whey were provided by the manufacturers. All media were sterilized by autoclaving at 120°C for about 15 min prior to their use for whey fermentation.

Materials Used as Carriers for Cell Immobilization

Corn Starch Gel

The CSG was prepared by gelatinization of 8 g corn flower with 100 mL deionized water at 90°C with agitation.

Delignified Cellulosic Material

The DCM was prepared from the wood sawdust treated by 1% w/v NaOH for 3 hours at the boiling point, in order for lignin to be removed and cavities to be created, enhancing their cell adsorption ability during immobilisation. Wood sawdust is the main waste of carpenter's shops, is an abundant, easy available and cheap material, popular for cell immobilization due to the relative ease of carrying out this by physical absorption.

Gluten Pellets

The GP were prepared as following: wheat flour was mixed with tab water, kneaded and the dough was washed with tab water in order to remove starch. The gluten remaining was shaped by hand into pellets of 1.5-2 cm diameter, which were dried at 105°C for 5 h. These pellets were cooled down to room temperature and used for cell immobilization.

Brewer's Spent Grains

In this study fresh BSG in their original form without any chemical treatment were used as immobilization supports. The kefir immobilization techniques on BSG are simple and handling is easy. The BSG are abundant, available at low or no cost throughout the year, are produced in huge quantities not only by large, but also by small breweries and their main application has been limited to animal feeding. The production of potable alcohol or whey alcoholic beverages using kefir immobilized on BSG seems to be a very attractive perspective.

Malt Spent Rootlets

The MSR in their original dried form were sieved and the particles with diameter less than 0.36 mm were selected and remained in water for two days. The wet MSR were used for kefir cells immobilization.

Methods

Cells Immobilization

For the preparation of CSG biocatalyst 6 g liquid biomass of kefir cells were added in the gelatinized corn flower, when temperature decreased at 40°C and for stabilization the mixture was remained for 24 h at 4°C. In all the other cases the cell immobilization on the various supports was carried out by suspending an amount of 16g of wet weight kefir co-culture in an 800 mL of synthetic culture medium containing 12% lactose and introducing 170g of sterilized support material. The systems were allowed to ferment for approximately 12 h at 30°C without agitation, then the fermented liquids were decanted and the immobilized biocatalysts were washed twice with 200 mL of fresh sterilized media and were used for the whey fermentations.

Fermentations

In order to study the effect of pH value and temperature on whey fermentation rate and lactose uptake rate by kefir cells entrapped in CSG, whey fermentations experiments were carried out in lab-scale 500-mL Erlemeyer flask without agitation or air supply. Specifically, the CSG biocatalyst was added in 250 mL of pasteurized whey and fermentations were carried out (i) at various pH values 4.5, 5.5 and 7 at 30°C and (ii) at various temperatures of 10, 20, and 30°C under 5.5 pH value. The experiments with free kefir cells or the other above described biocatalysts in this study were carried out at 30°C and under 5.5 pH value (except for DCM under pH 5) in 250 mL of pasteurized whey, where liquid biomass of 2.4% w/v or an amount of about 122.5 g immobilized biocatalyst was added. In all cases trial pH value was achieved by the addition of tartaric acid (7% w/v), as the original whey pH value was 6.5. During whey fermentation the pH value was maintained stable to the selected trial pH, by the addition of 6 M NaOH solution. At the beginning of each run a small quantity of ^{14}C-labelled lactose was added, in order to determine the lactose uptake rate by kefir during the processes. Samples of the fermented liquids were collected and stored at -20°C until further analysis. For statistical reasons all fermentations were carried out in triplicates.

^{14}C Labelled Lactose Determination

The labelled lactose [D-glucose-1-^{14}C], (ARC 0466 lactose, and 0.1 mCi/mL) was fermented in the same way as the non labelled one. At various time intervals samples of 5 mL were centrifuged in plastic vials and the gathered biomass was washed twice with deionized water centriguged again and weighted. The ^{14}C within the cells was determined by liquid scintillation. The amount of the labelled lactose consumed by the weighted amount of kefir biomass during fermentation was determined and expressed as counts per minute (cpm) of lactose per gram of kefir biomass per hour.

Statistical Methods

During each one of the whey fermentations, the standard deviation (SD) of every one of the three recorded values of residual sugars and ethanol concentrations as well as lactose uptake rates was calculated as follows:

$$s = \sqrt{\frac{1}{N-1}\sum_{i=1}^{N}(x_i - \overline{x})^2},$$

Where N is the number of repeats, (x_1, x_2 ...x_N) are the recorded values and is the mean value of them.

Assays

Liquid Scintillation Measurements

In the plastic vials with weighted kefir cells contained ^{14}C-labelled lactose 5 mL of liquid scintillation cocktail Opti Fluor (Perkin Elmer) was added and they were measured by liquid scintillation. All the measurements were performed on a PACARD-3255 liquid scintillation counter, interfaced to an APPLE-2 personal computer for data evaluation.

Determination of Residual Sugars and Ethanol

The kinetics of fermentation was monitored measuring Baume density (°Be) and residual sugars in the fermentation broth. Residual sugars in the whey samples were determined on a Shimadzu LC-9A HPLC system consisting of a Shim-pack SCR-101N column, an LC-9A pump, an RID-6A refractive index detector, a CTO-10A column oven, and a DGU-2A degassing unit. Three times distilled water was used as the mobile phase (0.8 mL/min), and 1-butanol (0.1% v/v) was used as an internal standard. Column temperature was 60°C. Sample dilution was 1% v/v, and injection volume was 40 μL (Tsaousi et al., 2008).

Ethanol was determined by gas chromatography using ID Hayeser Q cat#19017. Helium was used as carrier gas at 20 mL/min. Column temperature was programmed at 80-180°C at a rate 16°C/min. The temperatures of the injector and FID detector were 200 and 220°C, respectively. For the ethanol determination, a total volume of 1 μL for each sample was injected directly into the column and the concentration of ethanol was determined using standard curves. 1- Butanol was used as internal standard at a concentration of 0.5% (v/v) (Kopsahelis et al., 2009). Ethanol yield -the ratio of *g ethanol produced/g utilized sugar* in the fermentation- was calculated and presented in Table 1. Also, in Table 1 the lactose conversion calculated by the following equation is presented:

(Initial sugar conc. – Residual sugar conc.)/Initial sugar conc. × 100

The recorded values of residual sugar and ethanol concentrations were the mean value of three repeats and the standard deviation (SD) was calculated (Figures 4 and 6).

Electron Microscopy

Pieces of the CSG biocatalyst were washed with deionized water and dried overnight at 30°C. The samples were coated with gold in a Balzers SCD 004 Sputter Coater for 30 min and examined in a Jeol model JSM-6300 (Japan) scanning electron microscope (SEM).

Results

Kefir cells entrapment in CSG was proved by SEM (Figure 3) and by the efficiency of the resulted biocatalyst for the complete fermentation of the whey lactose at various temperatures and various pH values.

Specifically, the highest whey fermentation rate and lactose uptake rate by CSG biocatalyst was recorded at temperature of 30°C and at pH 5.5, as it was reported previously for kefir FC and GP, DCM, BSG and MSR biocatalysts (Golfinopoulos et al., 2009; Golfinopoulos et al., 2011; Golfinopoulos et al., 2012; Soupioni et al., in press).

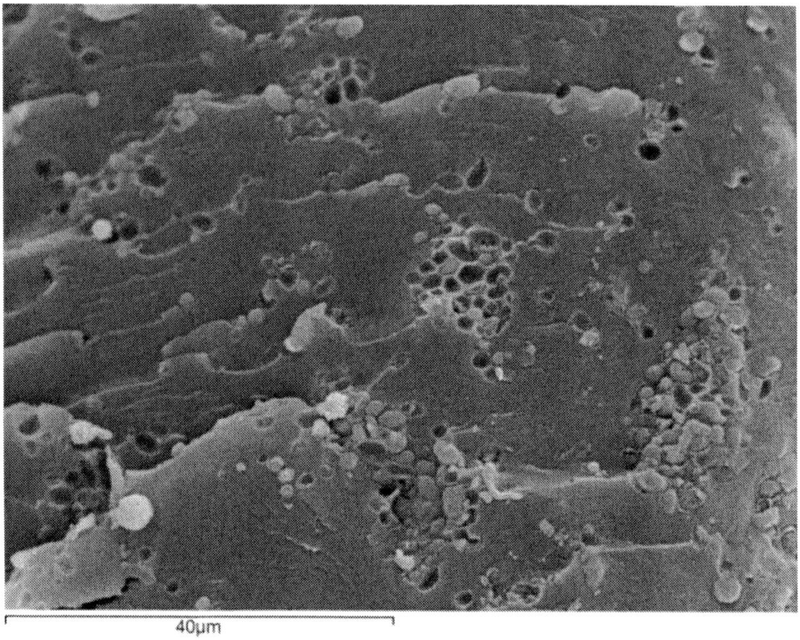

Figure 3. SEM micrograph showing the entrapped kefir cells in corn starch gel.

In Table 1 the whey lactose fermentation kinetics parameters by kefir FC, kefir cells immobilized on DCM, GP, BSG, MSR and entrapped in CSG at 30°C are presented. As in Table 1 and Figure 4 is illustrated the CSG biocatalyst was found suitable for lactose whey fermentation and impressively more effective in comparison with free cells. Moreover, Table 1 and Figure 4 show that the fermentation time -of only 10 hours- for CSG biocatalyst dropped about 37% in comparison with GP and MSR biocatalysts and about 86% compared with free cells. But the shortest whey fermentation time was for BSG and DCM biocatalysts, when the process lasted only about 8 hours and 7.2 hours respectively! It suggests that kefir cell immobilization on above materials acts as a promoter of the catalytic activity of the enzymes involved in the fermentation process, as previously reported (Bardi et al., 1996). Also, residual sugars were affected by the kind of support and were significantly low, ($P<0.05$), when DCM and BSG biocatalysts were used for whey fermentation (Table 1), in comparison with fermentations by GP, MSR and CSG biocatalysts.

Table 1. Kinetic parameters and average (N=3) concentrations (±SD) of residual sugars and ethanol produced during whey fermentations by kefir FC, kefir cells immobilized on DCM, GP, BSG, MSR and entrapped in CSG at 30°C

pH	Kind of biocatalyst	Temp (°C)	Initial Sugars (%w/v)	Fermentation Time (h)	Ethanol Concentration (%w/v)	Ethanol Yield (g/g)	Residual Sugars (%w/v)	Conversion (%)
5.5	**FC**	30	5.03	116.5	1.78±0.00	0.28	0.03±0.00	99.40
5.0	**DCM**	30	5.04	7.2	2.52±0.02	0.41	0.17±0.00	96.63
5.5	**GP**	30	5.[illegible]2	15	2.06±0.01	0.34	0.38±0.00	92.62
5.5	**BSG**	30	4.80	8	2.70±0. 03	0.46	0.20±0.00	95.83
5.5	**MSR**	30	5.[illegible]8	16	3.48±0.06	0.88	1.18±0.03	77.16
5.5	**CSG**	30	5.08	10	3.16±0.05	0.80	1.11±0.04	78.35

The results demonstrate the potential industrial applications mainly of fresh BSG in food fermentations with main features the low cost and simplicity of processing. BSG can be used as cell immobilization supports, without chemical treatment (e.g. delignification) which means easier handling and lower costs. Therefore, biocatalyst made by kefir cell immobilization on BSG can be acceptable by even small cheese-dairies as only a reactor with small volume is required to ferment the produced whey. Furthermore, due to the fact that BSG are industrial wastes produced in large quantities of negligible cost, their utilization is critical as a solution to their disposal problem. Eventually, using BSG biocatalyst could easily reduce the whey polluting load, maybe produce potable alcohol and a more nutritious animal feed.

The Figures 4 and 5 show that lactose uptake rate by kefir is strongly correlated to whey fermentation rate and increased during whey fermentation by kefir cells entrapped in CSG in comparison with fermentation by kefir FC. The same was earlier found for GP, DCM, BSG and MSR biocatalysts (Soupioni et al., in press).

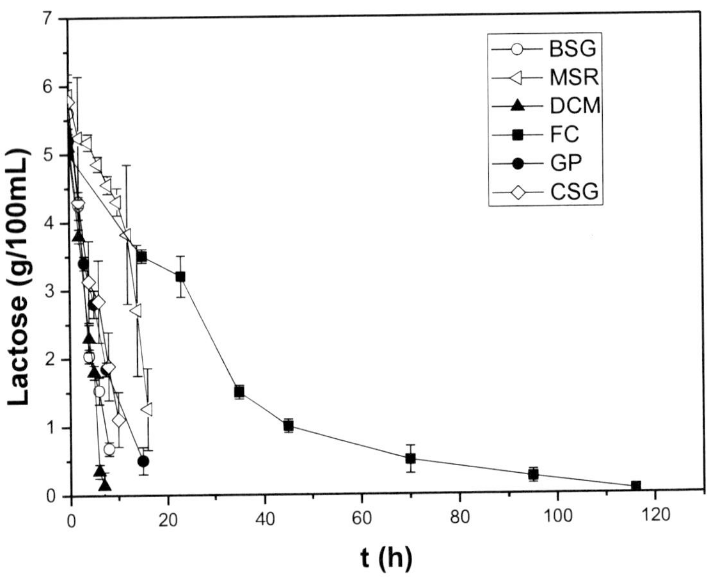

Figure 4. Fermentation kinetics of lactose by kefir free cells (FC) and immobilized ones on gluten pellets (GP), delignified cellulosic material (DCM), brewer's spent grains (BSG), malt spent rootlets (MSR) and starch gel (CSG) during whey fermentation at 30°C and 5.5 pH value (for DCM, pH = 5).

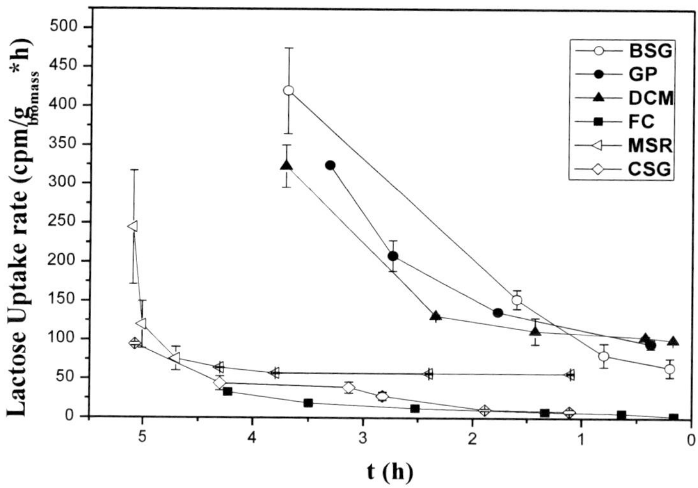

Figure 5. Lactose uptake rate by kefir free cells (FC) and immobilized ones on gluten pellets (GP), delignified cellulosic material (DCM), brewer's spent grains (BSG), malt spent rootlets (MSR) and starch gel (CSG) during whey fermentation at 30°C and 5.5 pH value (for DCM, pH = 5).

Furthermore, comparing BSG and DCM with GP and CSG is obvious that immobilization of kefir cells on cellulosic supports leads to higher fermentation rate and lactose uptake rate when compared to immobilization on no cellulosic ones.

In Figure 6, ethanol concentration during whey fermentation by kefir FC, immobilized on GP, DCM, BSG, MSR and entrapped in CSG is illustrated. The highest alcohol concentrations were obtained during whey fermentation by kefir cells immobilized on MSR and CSG, but the highest lactose uptake rates were obtained in fermentations by BSG and DCM biocatalysts. This implies that alcohol may be an inhibitor for lactic bacteria contained in kefir microflora, even at those concentrations -e.g. 3.16 g/100mL-, and therefore it plays an important role on the microorganisms fermentation activity and lactose uptake rate.

Ethanol concentration during whey fermentation by kefir cells immobilized on MSR reached near 3.48 g/100 mL whey. Distillation of the fermentation broth in order to obtain such a low ethanol concentration is very expensive. However, it is clear that the ethanol production in presence of kefir cells immobilized on MSR and CSG was about 2-fold higher compared with the ethanol obtained during whey fermentation by kefir FC. Therefore, further scientific research on cell immobilization is required in order to increase ethanol concentration during whey fermentation by kefir.

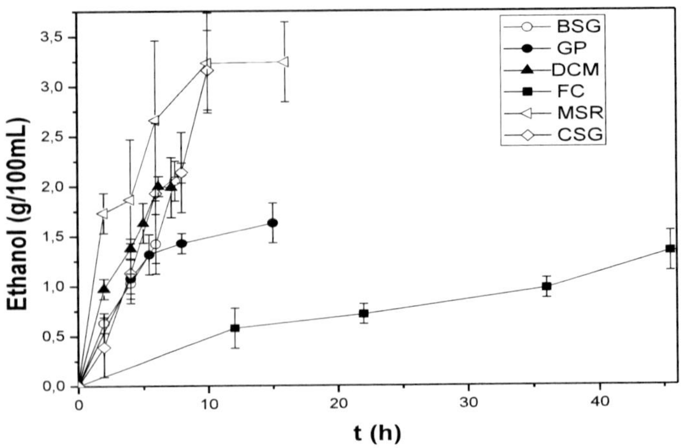

Figure 6. Ethanol concentration during whey fermentation by kefir free cells (FC) and immobilized ones on gluten pellets (GP), delignified cellulosic material (DCM), brewer's spent grains (BSG), malt spent rootlets (MSR) and starch gel (CSG) during whey fermentation at 30°C and 5.5 pH value (for DCM, pH = 5).

Conclusion

Biocatalysts prepared by immobilization of kefir cells on agro-industrial wastes such as BSG, MSR, CSG, GP and DCM were proved effective for whey fermentation at 30°C and pH 5.5, (for DCM, pH = 5), reducing significantly fermentation time and increasing lactose uptake rate in comparison to kefir FC. Specifically, using cheap, abundant and food grade BSG or DCM biocatalysts for whey fermentation is potentially acceptable by industries as the fermentation time was estimated only about 7 hours in that case. The highest alcohol concentration was observed by MSR biocatalyst.

Acknowledgments

I wish to thank my students Mrs. Sofia Maragou, MSc, Mrs. Foteini Vordou, MSc and Mr. Aris Golfinopoulos, PhD student for their valuable assistance during experiments. Also, I am grateful to my colleagues A.A. Koutinas and M. Kanellaki, Professors of Food Chemistry for their help.

References

Arasaratnam V., Senthuran A., Balasubramaniam K. (1996). Supplementation of whey with glucose and different nitrogen sources for lactic acid production by Lactobacillus delbrueckii. *Enzyme Microbiology Technology,* 19, 482-486.

Athanasiadis, I., Boskou, D., Kanellaki, M., Koutinas, A.A. (1999). Low temperature alcoholic fermentation by delignified cellulosic material supported cells of kefir yeast, *J. Agric. Food Chemistry*, 47, 4474-4477.

Athanasiadis, I., Boskou, D., Kanellaki, M., Koutinas, A.A. (2001). Effect of carbohydrate on fermentation by kefir yeast supported on delignified cellulosic materials. *Food Chemistry*, 49, 658-663.

Athanasiadis, I., Boskou D., Kanellaki, M., Kiosseoglou, V., Koutinas, A.A., (2002). Whey liquid waste of the dairy industry as raw material for potable alcohol production by kefir granules. *Food Chemistry*, 50, 7231-7234.

Athanasiadis, I., Koutinas, A.A., Bekatorou, A., Ikonomopoulou, M., Blekas, G. (2005). Kefir yeast technology: Scale- up in SCP production using milk whey, *Biotechnology and Bioengineering,* 89, 788-796.

Bardi E.P., Bakoyianis V., Koutinas A.A., Kanellaki M. (1996). Room Temperature and Low Temperature Wine Making Using Yeast Immobilized on Gluten Pellets. *Process Biochemistry*, 31(5), 425-430

Chaly E.A., Kamal A.M. (2004). Submerged yeast fermentation of acid cheese whey for protein production and pollution potential reduction. *Water Research*, 38, 631-634.

Chaly E.A., Kamal A.M., Correia R.L. (2005). Kinetic modelling of continuous submerged fermentation of cheese whey for single cell protein production. *Bioresource Technology*, 96, 1143-1152.

Dimitrellou, D., Kourkoutas, Y., Banat, I.M, Marchant, R., Koutinas, A.A. (2007). Whey-cheese production using freeze-dried kefir culture as a starter, *Journal of Applied Microbiology,* 103, 1170–1183.

Dragone G., Mussatto I.S., Oliveira M.J., Teixeira J. (2009). Characterization of volatile compounds in an alcoholic beverage produced by whey fermentation. *Food Chemistry*, 112, 929-935.

Garrote, G.L., Abraham, A.G., De Antoni, G.L. (1997). Preservation of kefir grains, a comparative study. *Lebensmittel–Wissenschaft un Technologie*, 30, 77–84.

Ghaly E.A., El-Taweel A.A. (1995). Effect of micro-aeration on the growth of *Candida pseudotropicalis* and production of ethanol during batch fermentation of cheese whey. *Bioresource Technology*, 52, 203-217.

Golfinopoulos, A., Papaioannou, L., Soupioni, M., & Koutinas, A.A. (2009). Lactose uptake rate by kefir yeast using ^{14}C-labelled lactose to explain kinetic aspects in its fermentation. *Bioresource Technology, 100,* 21, 5210-5213.

Golfinopoulos, A., Kopsahelis, N., Tsaousi, K., Koutinas, A.A., & Soupioni, M. (2011). Research perspectives and role of lactose uptake rate revealed by its study using 14C-labelled lactose in whey fermentation. *Bioresource Technology,* 102, 4204-4209.

Golfinopoulos, A., Soupioni, M., Kopsahelis, N., Tsaousi, K., Koutinas, A.A., (2012). Lactose uptake rate measurements by ^{14}C-labelled lactose reveals promotional activity of porous cellulose in whey fermentation by kefir yeast. *Food Chem.,* 134, 1973–1981.

Harta M.O., Iconomopoulou, M., Bekatorou, A., Nigan, P., Kontominas, M., Koutinas, A.A. (2004). Effect of various carbohydrates substrates on the production of kefir grains for the use as a novel baking starter. *Food Chemistry*, 88, 237-242.

Kandylis P., Goula A., Koutinas A.A. (2008). Corn Starch Gel for Yeast Cell Entrapment. A View for Catalysis of Alcoholic Fermentation. *Journal of Agricultural and Food Chemistry*, 56, 12037–12045.

Kargi F., Ozmihci S. (2006). Utilization of cheese whey powder (CWP) for ethanol fermentations: Effects of operating parameters. *Enzyme and Microbial Technology*, 38, 711-718.

Kopsahelis, N., Nisiotou, A., Kourkoutas, Y., Panas, P., Nychas, G.J.-E., Kanellaki, M., (2009). Molecular characterization and molasses fermentation performance of a wild yeast strain operating in an extremely wide temperature range. *Bioresource Technol.* 100, 4854-4862.

Kosseva, M.R., Panesar, P.S., Kaur, G., & Kennedy, J.F. (2009). Use of immobilized biocatalysts in the processing of cheese whey. *International Journal of Biological Macromolecules*, *45*, 437–447.

Koutinas A.A., Iconomopoulou M., Blekas G. (2005). Kefir yeast technology: Scale-Up in SCP production using milk whey. *Biotechnology and Bioengineering,* 89, 788-795.

Koutinas A.A., Athanasiadis I., Bekatorou A., Psarianos C., Kanellaki M., Agouridis N., Blekas G. (2007). Kefir-yeast technology: Industrial scale-up of alcoholic fermentation of whey promoted by raisin extracts. using kefir-yeast granular biomass. *Enzyme and Microbial Technology*, 41, 576-582.

Magalhaes K.T., Pereira G.V.M., Nicolau A., Dragone G., Dormingues L., Teixeira A., Silva B.J., Schwan R.F. (2010). Production of fermented whey-based beverage using kefir grains as starter culture: Evaluation of morphological and microbial variations. *Bioresource Technology*, 101, 8843-8850.

OECD-FAO Agricultural Outlook 2008–2017 Highlights. Paris: Organisation for Economic Co-operation and Development—Food and Agriculture Organization of the United Nations, 2008.

Ozmihci S., Kargi F. (2007). Effects of feed sugar concentration on continuous ethanol fermentation of cheese whey powder solution (CWP). *Enzyme and Microbial Technology*, 41, 876-880.

Ozmihci S., Kargi F. (2008). Ethanol production from cheese whey powder solution in a packed column bioreactor at different hydraulic residence times. *Biochemical Engineering Journal,* 42, 180-185.

Panesar, P.S., Kennedy, J.F., Gandhi, D.N. and Bunko, K. (2007). Bioutilisation of whey for lactic acid production. *Food Chemistry*, 105, 1-14.

Panesar S.P., Kennedy F.J., Knill J.C., Kosseva M. (2010). Production of L(+) lactic acid using *Lactobacillus casei* from whey. *Brazilian Archives of Biology and Technology.* 53, 219-226.

Paraskevopoulou A., Athanasiadis I., Blekas G., Koutinas A.A., Kanellaki M., Kiosseoglou V. (2003). Influence of polysaccharide addition on stability of a cheese whey kefir-milk mixture. *Food Hydrocolloids*. 17, 615-620.

Pescuma M., Hebert M.E., Mozzi F., Valdez F.G. (2010). Functional fermented whey-based beverage using lactic acid bacteria. *International Journal of Food Microbiology*, 141, 73-81.

Pintado, M.E., Lopes Da Silva, J.A., Fernandes, P.B., Malcata, F.X., Hogg, T.A. (1996). Microbiological and rheological studies on portugese kefir grains. *International Journal Food Science and Technology*, 31, 15–26.

Plessas, S., Pherson, L., Bekatorou, A., Nigam, P., Koutinas, A.A. (2005). Bread making using kefir grains as baker's yeast, *Food Chemistry,* **93**, 585–589.

Silveira B.W., Passos V.J.F., Mantovani C.H., Passos L.M.F. (2005). Ethanol production from chesse whey permeate by *Kluyveromyces marxi- anus*

UFV-3. A flux analysis of oxido-reductive metabolism as a function of lactose concentration and oxygen levels. *Enzyme and Microbial Technology*, 36, 930-936.

Samragy A.Y., Khorshid A.M., Foda I.M., Shehata E.A. (1996). Effect of fermentation conditions of production of citric acid from cheese whey by Aspergillus niger. *International Journal of Food Microbiology,* 29, 411-416.

Sansonetti S., Curcio S., Calabro V., Iorio G. (2009). Bio-ethanol production by fermentation of ricotta cheese whey as an effective alternative non-vegetable source. *Biomass and Bioenergy*, 33, 1687-1692.

Smithers, G.W. (2008). Whey and whey proteins-from 'gutter-to-gold. *International Dairy Journal*, 18, 695–704.

Soo J.K., So S.J., Heo R.T. (1994). Alcohol production from whey in batch and continuous culture of *Kluyveromyces fragilis*. *Journal of Microbiology and Biotechnology*, 4(4), 333-337.

Soupioni, M., Polichroniadou, E., Tokatlidou, M., Kanellaki, M., Koutinas, A.A., (1998). Glucose uptake rate by *Saccharomyces cerevisiae* in the presence of promoters of alcoholic fermentation using ^{14}C-labelled glucose. *Biotechnology Letters*, 20 (5), 495-497.

Soupioni, M., Golfinopoulos, A., Kanellaki, M. and Koutinas, A.A., Study of whey fermentation by kefir immobilized on low cost supports using ^{14}C-labelled lactose. *Bioresource Technology,* doi:10.1016/j. biotech. 2012. 12.131 (in press).

Tsaousi, K., Dimitrellou, D., & Koutinas, A.A. (2008). Low-temperature thermal drying of Saccharomyces cerevisiae starter culture for food production. *Food Chemistry,* 110, 547-553.

Zafar S., Owais M. (2006). Ethanol production from crude whey by *Kluyveromuces marxianus*. *Biochemical Engineering Journal*, 27, 295-298.

Zaied A.K., Kawther S.K., Essawy E.A., Darwish A.A. (2009). Enhancement the fermentation of lactose-whey using recombinants between *Lactobacillus* and *Tetragenococcus*. *Australian Journal of Basic and Applied Sciences*, 3(2), 761-775.

In: Lactose
Editors: David Green and Emma Lee
ISBN: 978-1-62808-198-5

Chapter 4

Crystalline and Amorphous Lactoses: Tableting Properties and Their Application for the Production of Pharmaceutical Compacts

John Rojas, Santiago Hernandez and Alejandro Giraldo
Department of Pharmacy, School of Pharmaceutical Chemistry, University of Antioquia

Abstract

Lactose is the most widely used excipient for the manufacture of tablets and capsules due to its relatively low cost, high solubility, good absorption and rapid metabolism in the human body. It is blended with a drug and other excipients and compressed to make tablets for oral administration. However, lactose is not desirable to formulate alkaline drugs or active ingredients with amine moieties due to the formation of a Maillard reaction. It is available as α-lactose monohydrate, anhydrous α-lactose, anhydrous β-lactose and spray-dried lactose.

Spray-dried lactose had the best flow. On the contrary, β-lactose exhibited the poorest flow, but along with the anhydrous α-lactose presented the highest binding ability. Conversely, α-lactose monohydrate presented the poorest compactibility, the fastest disintegration time and its compactibility was not affected by moisture.

Spray-dried lactose is composed of α-lactose monohydrate crystals glued together by amorphous lactose. On the contrary, α-lactose and α/β-lactose were the most fragmenting materials. In turn, these materials were the least sensitive to magnesium stearate. Lactose tablets independent of the crystalline form showed a moisture-dependent tensile strength resulting into a decreased compactibility. However, under storage compacts showed a very small elastic recovery.

It is concluded that compactibility and compressibility of lactose was related to its polymorphic form.

Introduction

Currently, more than 80% of all dosage forms on the market are comprised of tablets because they are easy to dispense, offer dosage accuracy and present better stability to heat and moisture compared to liquid and semi-solid formulations. These dosage forms are mainly produced by wet granulation (WG), and direct compression (DC). Direct compression is a simple and economical method by which compacts are made directly from a powder blend of drug and excipients. The main advantage of DC over other tablet manufacturing methods is its simplicity, since it requires few unit operations and utilizes much less, energy, making the process more economical. On the other hand, in wet granulation, the drug is mixed with a wet binder and other excipients such lactose and passed through sieves to obtain granules suitable for tableting. Typically, a tablet is composed of several specialized excipients, such as a binder (compact forming material), filler (diluent), a disintegrant (facilitates compact disintegration) and a glidant/lubricant (improves powder flow and reduces friction with punch-die tooling).

Polymorphism occurs when the same material exhibits different crystal structures within the solid state, but has identical characteristic in solution. Lactose, which is the most widely used excipient in tablet formulation, especially by employing the process of wet granulation, occurs in different crystalline forms. The same type of lactose can be processed under different conditions rendering α and β-lactoses or their mixture (Crisp et al., 2007). α-lactose monohydrate is produced by crystallization from a supersaturated solution at temperatures below 93°C, whereas β-lactose is obtained at temperatures above 93°C (Crisp et al., 2010). The proportion of α and β-lactose produced depends on the crystallization conditions. During crystallization of β-lactose, no water is incorporated in the crystal lattice and thus, it exists only in the anhydrous form. In contrast, α-lactose exist in the hydrous and anhydrous form (Haque and Roos 2005). Dehydration with methanol converts α-lactose monohydrate into a stable anhydrous form, whereas a thermal treatment above 130°C renders a non-stable anhydrous α-lactose (Listiohadia et al., 2008).

In order to ensure a robust and successful manufacture of tablets, an excipient ideally should possess the following characteristics: excellent compressibility, adequate powder flow, resistance to segregation during handling and storage, fast compact disintegration, a broad range of bulk densities, low sensitivity to lubricants, and it should not have a complex production. The goal of this study is to assess the effect of the polymorphic form on the functional properties of lactose and compare those to commercial spray-dried lactose.

Materials and Methods

Materials

α-Lactose monohydrate (lot 01-20-187-2) and spray-dried lactose (lot 85120810620 were obtained from DmV international (Veghel, Netherlands) and Foremost Farms (Baraboo, WI, USA). Magnesium stearate (lot 2256KXDS) and ethylenglycol (lot 5001Y12D53) were obtained from Mallinckrodt Specialty Chemicals Co (St. Louis, MO, USA). The fraction of lactose materials that passed through a 200 sieve mesh was used for the study.

Production of Anhydrous α-Lactose

Approximately, 50g of α-lactose monohydrate was dispersed in 100 mL of ethylenglycol and heated at 150°C until a complete solution was formed. The heating was continued until reaching a volume of 50 mL. This solution was then poured on petri dishes and dried at 110°C until the α-Lactose crystals were obtained. These crystals were then cooled down at room temperature, washed with 250 mL of n-propanol and dried again at 105°C for 3h. This material was passed through a 200 mesh sieve before employed.

Production of the α/β-Lactose (40:60) Mixture

Approximately, 50g of α-lactose monohydrate was dispersed in 200 mL of distilled water and heated at 95°C until a complete solution was formed. The heating was continued until a complete evaporation of the solvent took place. The resulting mass was then dried on an oven at 110°C for 3h. This material was passed through a 200 mesh sieve before used.

Production of β-Lactose

Approximately, 50g of α-lactose monohydrate was dispersed in 100 mL of distilled water and heated at 105°C until dryness. This material was kept on an oven (U50, Memmert, Germany) at 110°C for 24 h. The dry material was passed through a 200 mesh sieve before employed.

Fourier-Transform Infrared Spectroscopy (FT-IR)

Approximately 1.5 mg of sample was mixed with about 300 mg of dry potassium bromide (dried at 110 °C for 4 hours before use) with an agate mortar and pestle. The powdered sample was compressed into a pellet using a 13 mm flat-faced punch set and a die on a manual press at ~160 MPa. The dwell time was 30 seconds. A Perkin Elmer FT-IR spectrophotometer (Spectrum BX, Perkin Elmer, CA), equipped with Ommic® software (Nicolet Corp, Madison, WI), was used to obtain the spectrum between 400 to 950 cm^{-1} using an 8 resolution and 16 scans.

Powder X-Ray Difractograms (PXRD)

Powder X-ray diffractions (PXRD) measurements were conducted over a 5 to 40° 2θ range using a Rigaku diffractometer (Model Miniflex II, Rigaku Americas, The woodlands, TX) at 40 kV and 30 mA, equipped with a monochromatic CuKα (α_1 = 1.5460 Å, α_2 = 1.54438 Å) X-ray radiation. The sweep speed and step width were 0.5° 2θ/min and 0.008°, respectively. The PeakFit software (v. 4.12, Seasolve software, Inc, Framingham, MA) was used to analyze the data.

Powder Properties

The optical microphotographs were taken on an optical microscope (BM-180, Boeco, Germany) coupled with a digital camera (DCR-SX45, Sony, Tokyo, Japan) at a ×2800 magnification. Particle size expressed as the geometric mean diameter was obtained from the microphotographs taken at a x70 magnification. Two hundred particles were randomly selected in each picture and their projected diameter (d_p) was calculated using the ImageJ software (v. 1.37, National Institute of Health, Bethesda, MD) according to the relationship:

$$d_p = 2 * \frac{A}{\pi}^{0.5} \tag{1}$$

The log-normal distribution plot was then constructed between d_p and the cumulative percent frequency to calculate the geometric mean diameter.

The bulk density was determined by taking the ratio of 3.5 g of sample divided by the measured volume taken on a 10 mL graduated cylinder. The tap density was measured directly from the final volume of the tapped sample obtained from the Auto-Tap analyzer (AT-2, Quantachrome instruments, USA) operated for 500 taps.

The true density was determined on a helium picnometer (AccuPycII 1340, Micromeritics Corp, USA) with ~3 g of sample. Powder porosity (ε) was determined from the equation:

$$\varepsilon = 1 - \frac{\rho_{bulk}}{\rho_{true}} * 100\% \tag{2}$$

where, ρ_{bulk} and ρ_{true} correspond to the bulk and true densities, respectively.

The compressibility of the powder was obtained by applying the Kawakita model (Kawakita and Ludde, 1971):

$$N/C = P/a + 1/ab \quad C = (V_i - V_n)/V_i \tag{3}$$

where, N is the tap number, V_i the initial volume and V_n the volume at the respective tap number.

The constant "a" is related to the total volume reduction for the powder bed (compressibility index) and the constant "b" is related to the resistant forces (friction/cohesion) to compression.

Flowability, expressed as flow rate (g/s) was determined by filling a glass funnel (13 mm hole diameter) with ~3.5 g of powder and measuring the flow time through the orifice. Moisture content was determined on an Ohaus moisture analyzer (MD 200, UK) loaded with ~3g of sample and operated for 15 min.

Preparation of Compacts

Compacts of ~500 mg were manufactured on a single station tablet press (Compac 060804, Indemec, Columbia) equipped with a 13-mm flat-faced punches and die tooling at compression pressures from 15 to 200 MPa and a dwell time of 1 s.

The upper punch was equipped with a load cell (LCGD-10K, Omega Engineering, Inc, Stanford, CT) and a strain gauge meter (DPiS8-EI, Omega Engineering, Inc, Stanford, CT).

Compressibility Analysis

Compact volume was calculated using the relationship: $V = h\,\pi r^2$, where r (mm) is the radius and h is the thickness (mm) of the compact. Compact dimensions (diameter and thickness) were measured with an electronic caliper (Titex Gold, Impofer, 0.01mm sensitivity). Compact densification data were fitted to the Heckel model (Heckel, 1961):

$$Ln(-\varepsilon) = kP + A \tag{4}$$

where, ε is the porosity of the compact at a pressure P and A the intercept, which represents the powder densification by die filling and particle rearrangement before particle deformation and bonding take place. The slope (k) of the linear portion of the plot is inversely related to the mean yield pressure (P_y), which is a measurement of the plasticity of the compressed material. On the other hand, D_a is related to the total compact densification and was obtained from the intercept (A) by the expression:

$$D_a = 1 - \exp^{-A} \qquad (5)$$

Compact Tensile Strength and Compactibility

It was determined on a VanKel hardness tester (UK 200, VanKel, Manasquan, NJ, USA). Each compact was placed between the platens and the crushing force was then measured.

The radial tensile strength (TS) values were obtained according to the Fell and Newton (Fell and Newton, 1968) method using the crushing force data. The area under curve of the tensile strength (AUCTS) was used to determine the material compactibility.

Compact Disintegration

Tablets, each weighing 500 mg, were made on a single punch tablet press using a 13 mm round, flat-faced punches and die set at a compression forces ranging from ~15-200 MPa and a dwell time of 1 s. Compacts were tested in distilled water at 37°C employing a disintegration apparatus without discs (39-133-115, Hanson Research Corporation, Northridge, California, USA) operated at 30 strokes/min.

Lubricant Sensitivity

It was assessed by mixing lactose powders with magnesium stearate at a 99:1weight ratio for 15 min. Tablets (~500 mg) were prepared using a single punch tablet press at 125 MPa and at a dwell time of 1 s. Lubricant sensitivity was expressed as the lubricant sensitivity ratio (LSR):

$$LSR = (H_0 - H_{lub})/H_0 \qquad (5)$$

where, H_0 and H_{lub} are the tensile strengths of tablets prepared without and with lubricant, respectively.

Reprocessing Susceptibility

Round compacts of ~500 g were made on a single punch tablet press using a 13 mm round flat-faced tooling, a dwell time of 1 s and a compression pressure of 125 MPa. Compacts were analyzed for tensile strength as described under "compact tensile strength" and pulverized on a mortar and pestle. The powder was passed freely on a 200 mesh sieve, compressed again under the same conditions, and the resulting tensile strength was determined.

Compact Elastic Recovery

Compacts were made as described under "Preparation of tablets" at a compression pressure of 125 MPa. Compact heights were measured with a caliper (Titex Gold, Impofer, 0.01mm sensitivity) once ejected and stored in a desiccator (RH = 15–30%) and after three days compacts heights were measured again.

The compact elastic recovery was expressed as percentage and calculated from the Armstrong and Haines-Nutt equation as reported previously (Armstrong and Haines, 1972).

Results and Discussion

Figure 1 shows the PXRD of the different polymorphs obtained. Alpha lactose monohydrate showed the characteristic crystalline peaks at 19.2°, 19.6°, 20.1° and 21.3°. Anhydrous α-lactose exhibited the main crystalline peaks at 18.2°, 18.8°, 19.4°, 19.9° and 20.8°. This material also showed some amorphous component due to the ethylenglycol used for recrystallization, which in turn was removed by washings with n-propanol. Commercial spray-dried lactose also showed the characteristic peaks of α-lactose monohydrate and one new peak at 16.5 which has been attributed to an impurity in certain lactose products (Miao S. and Ross 2005). Beta lactose exhibited the characteristic crystalline peaks at 10.9°, 19.5° and 21.3° (Thomas et al., 2009). In this case, the peak at 10.9° was very intense and the peak at 20.1 was virtually absent. Likewise, the material composed of a mixture of α/β lactose presented peaks at 10.9° and 20.1° of low intensity.

The crystallization conditions at 95°C favored the formation of lactose as a 40α:60β mixture. Only recrystallization at temperatures higher than 100°C for 24 h promoted the formation pure β-lactose. It has been reported that β-lactose can also be obtained from alcoholic potassium methoxide solutions, although the level of purity depends on the experimental conditions (Kirk et al., 2007; Farber and Tardos, 2003).

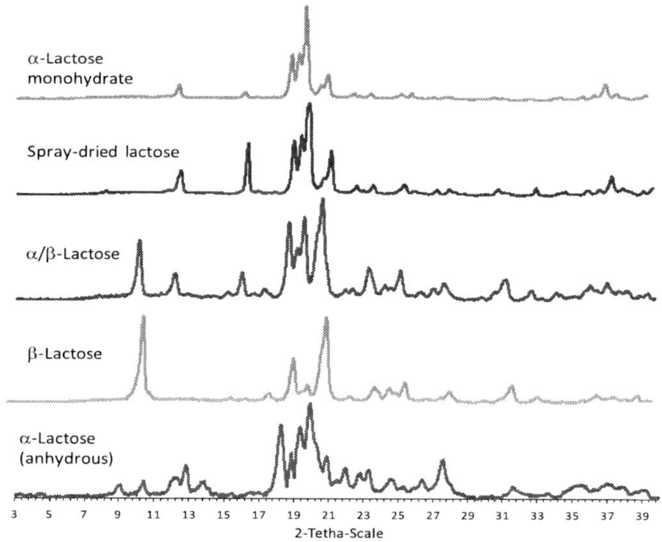

Figure 1. PXRD of different lactose allomorphs.

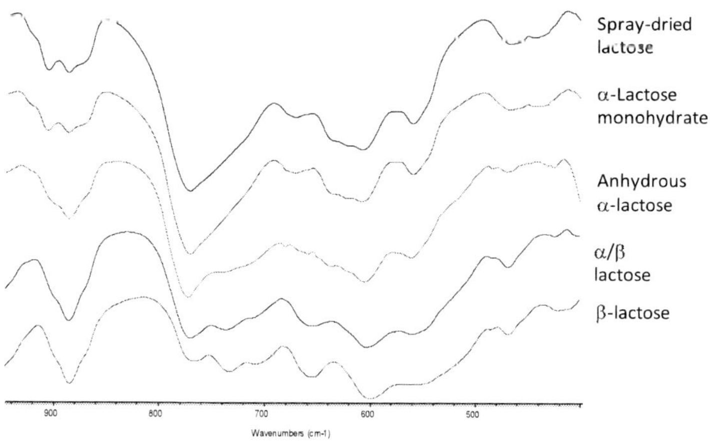

Figure 2. FT-IR spectra of different lactose allomorphs.

Figure 2 shows the FT-IR spectra of the lactose allomorphs in the fingerprint region, which was the only region that truly differentiated the lactose polymorphs. Beta lactose and the α/β mixture showed a characteristic band at 885cm^{-1}, whereas materials containing α-lactose in some degree showed a band at 770 cm^{-1}.

The morphology of the lactose particles is shown in Figure 3 and depended on the recrystallization and milling conditions. For instance, anhydrous α-lactose which was recrystallized in ethylenglycol at high temperatures showed irregular and elongated crystals. Other materials exhibited aggregates of irregular shape. Further, the semispherical shape of spray-dried lactose was not affected by the sieving process conducted through a 200 mesh and was responsible for the fast flow rate shown by this material. Commercial spray-dried lactose is produced by spray drying a suspension of α-lactose monohydrate crystals. This process results in a product composed of α-lactose monohydrate crystals glued together with amorphous lactose. However, the PXRD did not show a high amorphous component due to the recrystallization of the amorphous part back into α-lactose monohydrate crystals. It is possible that this conversion occurs during the shelf life of spray-dried lactose and is fastened if stored at high relative humidities.

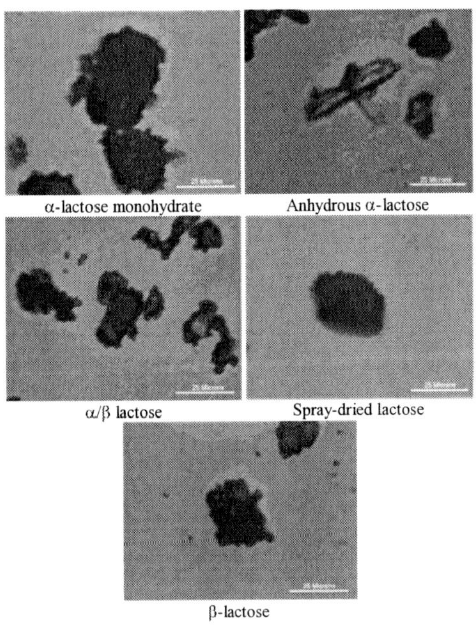

Figure 3. Optical microphotographs of different lactose allomorphs.

Table 1. Powder and compaction properties of several lactose allomorphs

Property	α-lactose monohydrate	α/β-lactose	Spray-dried	Anhydrous α-lactose	β-lactose
Powder porosity	0.74± 0.07	0.72±0.04	0.69 ± 0.04	0.64± 0.19	0.69±0.10
Geometric mean diameter (μm)	23.2	20.3	27.3	25.6	24.3
Flow rate (g/s)	5.9	6.2	8.1	3.1	3.8
Moisture content (%)	2.5	2.2	2.8	0.8	0.6
P_y	149	136.7	126	127.1	126.5
D_A	0.68	0.75	0.67	0.66	0.74
AUCTS	52.9	172.9	251.3	432.6	279.1
a	0.85	0.86	0.83	0.79	0.83
b	0.6	0.75	0.73	0.51	0.22
Lubricant sensitivity ratio	0.09	0.06	0.10	0.34	0.09
Elastic recovery (%)	0.54 ± 0.18	0.24 ± 0.21	0.23 ± 0.20	0.73 ± 0.37	0.46± 0.2

Since all materials were passed through a 200 mesh sieve and only the fraction retained between 200-325 mesh was used, the particle size ranged between 20.3 and 27.3 μm. Thus, differences among the properties studied can only be attributed to the crystalline form of lactose. Anhydrous α-lactose was the least porous material due to the elongated and irregular structure of the crystals which allowed for the formation of a highly packed powder bed.

Compressibility is expressed as the relative volume reduction of the powder bed in response to the applied pressure, and compactibility is the ability to form a compact with sufficient strength when a compression force is applied. Materials, by virtue of their response to applied forces, can be classified as elastic, plastic, or brittle materials. Generally, materials are not entirely elastic, plastic or brittle, but have some of all three characteristics. Thus, materials may exhibit all three types of behavior, with one type being the predominant response. Opposed to brittle materials, a ductile material can withstand large deformations without breaking. Brittleness is caused by progressive failure along weak points in the crystals, whereas ductility favors sliding of crystal planes. Ductility and brittleness favor bonding because new surfaces are produced and, consequently, an increase in contact area between particles occurs during compression. Figure 4 shows the Heckel plots of the lactose allomorphs. These plots describe the type of deformation upon compression. The Heckel plots showed that compact densification increased with increasing forces. Alpha lactose monohydrate and α/β lactose presented the highest powder yield pressures indicating that these were the most fragmenting materials upon consolidation in the powder bed. Conversely, spray-dried, anhydrous α-lactose and β-lactose were the least fragmenting materials. On the other hand, the volume reduction ability of the lactose powders expressed as the "a" compressibility index showed a narrow range from 0.79 to 0.86 and was mainly dependant on the particle size, which in turn was very similar in all cases.

Compacts made of anhydrous α-lactose and β-lactose presented the highest tensile strength profiles. Conversely, α-lactose monohydrate rendered the weakest compacts. The area under the curve of the tensile strength indicated that anhydrous α-lactose and β-lactose were the most compactable materials. Results indicate that it is beneficial to have anhydrous α-lactose and β-lactose for making compacts by direct compression, whereas α-lactose monohydrate and other lactose allomorphs are only recommended for the process of wet granulation. Hydrophobic lubricants and especially magnesium stearate has a deteriorating effect on the compactibility of the excipients. Magnesium stearate interferes bonding by the formation of a lubricant film around particles during the process of dry blending. Fragmenting particles are less sensitive to magnesium stearate as compared to plastically deforming excipients such as microcrystalline cellulose and starch. This is attributed to the creation of lubricant free surfaces on the particles.

Therefore, sensitivity to magnesium stearate can be used to assess the fragmentation propensity of lactose materials. Thus, lubricant sensitivity should decrease with increasing degree of fragmentation. Results indicate that a highly compactable material with a low fragmenting character such as anhydrous α-lactose had the highest sensitivity to magnesium stearate. This is due to the low ability of anhydrous α-lactose to generate magnesium stearate free surfaces upon consolidation resulting in a low binding capacity. Fragmentation of lactose particles during consolidation might occurs within the areas surrounded by the magnesium stearate network. Thus, lubricated compacts failed along the interfaces of the original particles.

The elastic recovery of lactose materials was very low (<1%) indicating a good behavior after compaction since the applied stored energy was not release upon compact ejection. This behavior is typical of fragmenting materials.

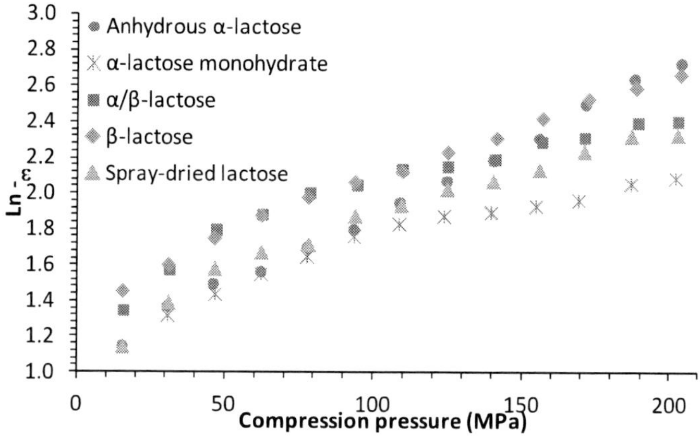

Figure 4. Heckel plots of lactose allomorphs.

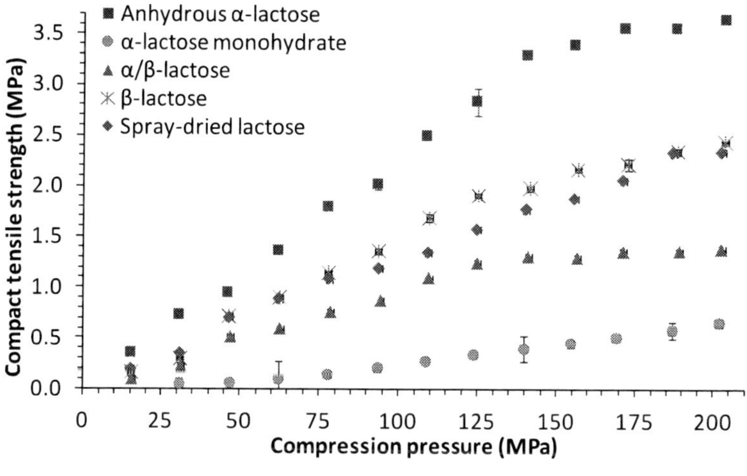

Figure 5. Compact tensile strength of lactose allomorphs.

Materials that rendered the strongest compacts such as anhydrous α-lactose exhibited the slowest disintegrating compacts and showed a linear profile (Figure 6). On the contrary, α-lactose monohydrate, which generated the weakest compacts showed a rapid disintegration upon water contact and its disintegration profile was independent of the compact porosity. Spray-dried lactose and β-lactose presented similar profiles due to their comparable compactibility. The binding capacity of lactoses decreased with increasing water content (Figure 7). The least affected material was α-lactose monohydrate, whereas spray-dried lactose was the most affected, probably due to its amorphous regions that have a high affinity for water. On the other hand, anhydrous α-lactose could pick up water rapidly becoming α-lactose monohydrate decreasing its compactibility. The presence of water may lead to preferred orientations in particle bonding, non-homogeneous particle slippage and differences in pressure transmission within the powder bed, all possibly resulting in tablet lamination and capping. Lamination corresponds to the separation of a compact into two or more distinct horizontal layers, whereas capping occurs when the upper or lower segment of the compact separates horizontally from the main body.

It is important to know the reprocessing behavior of a material to determine its likelihood for dry granulation in a pharmaceutical process. Sometimes after milling a material could lose its ability for the formation of compacts because the number of binding points is reduced. Results show that the binding ability of the materials remained unchanged except for α-lactose monohydrate and spray-dried lactose (Figure 8). The increase in compactibility of α-lactose monohydrate could be due to the reduction in porosity upon reprocessing.

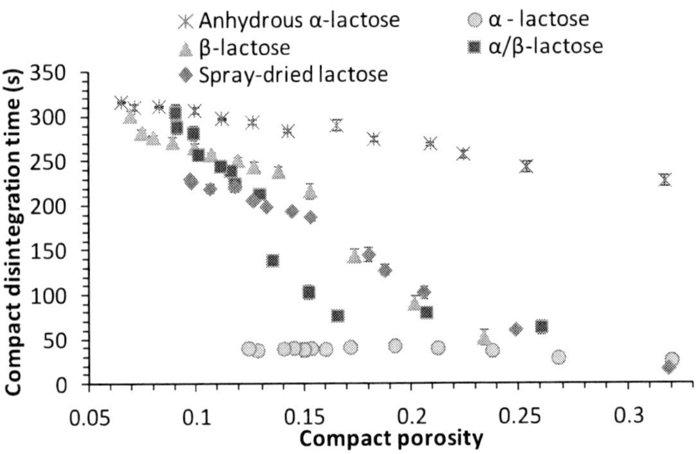

Figure 6. Compact disintegration time of lactose allomorphs.

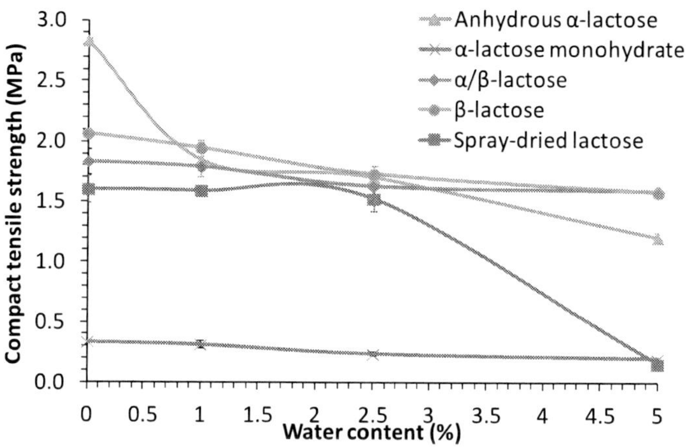

Figure 7. Compact susceptibility to water content.

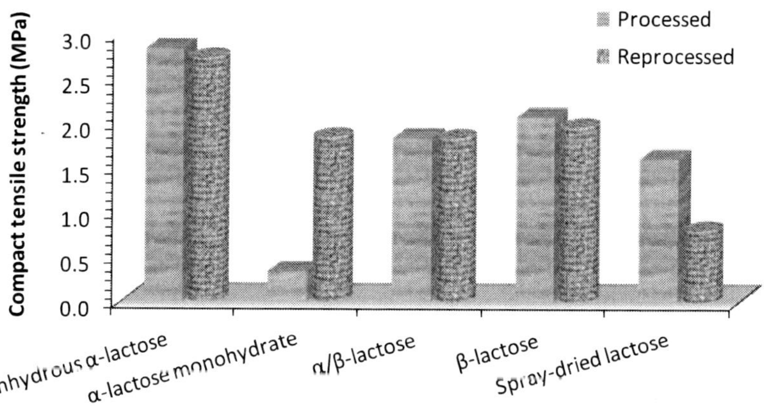

Figure 8. Susceptibility of lactose allomorphs to reprocessing.

Conclusion

All lactose allomorphs consolidate by particle fragmentation and α-lactose and α/β lactose were the most fragmenting materials. The strength of lactose particles is attributed to Van der Waals forces and in a less degree by hydrogen bonding. Lactose compacts disintegrated by dissolution. The disintegration characteristics depended on the binding properties of the crystalline forms.

Acknowledgments

The authors thank the committee for the development of research CODI and the Pharmaceutical Sciences institute of the University of Antioquia for providing us with the resources needed to conduct this project.

References

Armstrong NA, Haines-Nutt RF (1972). Elastic recovery and surface-area changes in compacted powder systems. *J. Pharm. Pharmacol.,* 24:135–136.

Crisp JL, Dann SE and Blatchford CG (2007). Antisolvent precipitation of lactose from aqueous solutions. *Int. J. Pharm.,* 334:103-114.

Crisp JL, Dann SE, Edgar M, Blatchford CG (2010). The effect of particle size on the dehydration/rehydration behaviour of lactose. *Int. J. Pharm.,* 391:38–47.

Farber L, Tardos GI (2003). Evolution and structure of drying material bridges of pharmaceutical excipients: studies on a microscope slide. *Chem. Eng. Sci.,* 58: 4515-4525.

Kawakita K and Ludde KH (1971). Some considerations on powder compression equations. *Powder Technol.,* 4:61-68.

Fell JT and Newton JM (1968). The tensile strength of lactose tablets. *J. Pharm. Pharmacol.,* 20:657-758.

Haque MK and Roos YH (2005). Crystallization and X-ray diffraction of spray-dried and freeze-dried amorphous lactose. *Carbohyd. Res.,* 340: 293–301.

Heckel RW (1961). An analysis of powder compaction phenomena. *Trans Metal. Sci. AIME.,* 221:1001-1008.

Kirk JH, Danna SE, Blatchford CG (2007). Lactose: A definitive guide to polymorph determination. *Int. J. Pharm.,* 334: 103-114.

Listiohadia Y, Hourigana JA, Sleighb RW, Steeleb RJ (2008). Moisture sorption, compressibility and caking of lactose polymorphs. *Int. J. Pharm.,* 359: 123–134.

Miao S. and Ross Y (2005). Crystallization kinetics and X-rays diffraction of crystals formed in amorphous lactose, trehalose and lactose/trehalose mixtures. *J. Food. Sci.,* 70: 350-358.

Thomas NR, Shumway LS, Lee DH (2009). Quantitative X-Ray Diffraction. Determination of lactose monohydrate and "lactose in chocolate". *J. Food. Sci.,* 74: 513-518.

In: Lactose
Editors: David Green and Emma Lee

ISBN: 978-1-62808-198-5

Chapter 5

Lactose and its Influence in Diffusion of Divalent Metal Chlorides

Ana C. F. Ribeiro[1*], Luis M. P. Verissimo[1], Victor M. M. Lobo[1] and Miguel A. Esteso[2†]

[1]Department of Chemistry, University of Coimbra, Coimbra, Portugal

[2]Departamento de Química I, U.D. Química Física, Facultad de Farmacia, Universidad de Alcalá, Alcalá de Henares, Madrid, Spain

Abstract

Lactose (disaccharide derived from the condensation of galactose and glucose) is not only a technological important compound, but it also enjoys biological, medical, pharmaceutical, food and biomedical applications [1-4]. As a typical non-electrolyte carrying hydrophilic hydroxyl groups capable of hydrogen-bonding, its properties play a significant role in the reaction conditions of many current industrial processes, such as enzymatic conversion of biomass to useful chemicals, and in many biochemical interactions with metal ions.

---

* Corresponding author: Tel: +351-239-854460; Fax: +351-239-827703; Email: anacfrib@ci.uc.pt luisve@gmail.com.

† Email: miguel.esteso@uah.es.

However, while numerous studies have been carried out on the thermodynamic properties of binary aqueous lactose solutions, data are more limited on the transport behaviour of these sugar systems in aqueous solutions. The characterization of the transport properties, particularly diffusion coefficients, is important for fundamental reasons, helping us to understand the nature of the structure of the aqueous systems containing metal ions, and for practical application in fields such as corrosion. In fact, they provide a direct measure of molecular mobility, an important factor in the preservation of biological materials in sugar matrixes.

Lactose is the most important carbohydrate present in the milk of most mammals. As a consequence, for the lactose-intolerant persons it has been verified that the ingestion of enough calcium may be difficult. For these people, who reveal inability or insufficient ability to digest lactose, because of a deficiency of the enzyme lactase, it is often recommended that they have calcium supplements, such as calcium carbonate or others.

Hopefully, the studies reported here will lead to an increase in the know-how, which will allow a better understanding of the physicochemical conditions underlining the diffusion phenomena occurring in different biological systems. Thus, we proposed to extend our studies concerning the experimental determination of the diffusion coefficients for systems containing lactose [5-9], that is, for aqueous system of calcium chloride plus lactose, having in mind a better understanding of the structure of these systems, and of the thermodynamic and transport behaviour of this carbohydrate in those aqueous media. On the other words, the main scope of our work shown here is to show the effect of lactose on the diffusion of metal ions involved in biological systems.

Keywords: Diffusion Coefficients; Aqueous Solutions; Electrolytes; lactose

1. Introduction

1.1. Concepts of Diffusion

The gradient of concentration inside the solution (without convection or migration) produces a flow of matter in opposite direction, which arises from random fluctuations in the positions of molecules in the space. This phenomenon, denominated diffusion, is an irreversible process. In fact, the gradient of chemical potential in the real solution is treated as the true virtual force producing diffusion. However, in ideal solutions, that force can be quantified by the gradient of the concentration at constant temperature. Thus, we may consider the following approaches to describe the isothermal diffusion: the thermodynamics of irreversible processes and Fick´s laws [10, 11].

Diffusion coefficient, *D,* in a binary system, may be defined in terms of the concentration gradient by a phenomenological relationship, known as Fick´s first law,

$$J = -D\frac{\partial c}{\partial x} \tag{1.1}$$

where *J* represents the flow of matter across a suitable chosen reference plane per area unit and per time unit, in a one-dimensional system, and *c* is the concentration of solute in moles per volume unit at the point considered; (1.1) may be used to measure *D*. The diffusion coefficient may also measured considering Fick's second law

$$\frac{\partial c}{\partial t} = \frac{\partial}{\partial x}\left(D\frac{\partial c}{\partial x} \right) \tag{1.2}$$

In general, the available methods are grouped into two groups: steady- and unsteady-state methods, according to (1.1) and (1.2). In most of the processes, diffusion is a three-dimensional phenomenon. However, many of the experimental methods used to analyze diffusion restrict it to a one dimensional process and, also, it is much easier to study its mathematical treatments in one dimension (which may be afterwards generalized to a three-dimensional space).

The resolution of equation (1.2) for a unidimensional process is much easier if we consider *D* as a constant. This approximation is applicable only when there are small differences of concentration, which is the case in our open-ended condutimetric technique [12], and in the Taylor technique [13]. Under these conditions, it is legitimate to consider that our measurements of differential diffusion coefficients obtained by the above techniques are parameters with a well defined thermodynamic meaning [10-13].

1.2. Measurement of Diffusion Coefficients in Binary, Pseudo-binary and Ternary Systems: Conductimetric and Taylor Dispersion Techniques

The open-ended capillary cell used, which has previously been applied to obtain mutual diffusion coefficients for a wide variety of electrolytes (e.g.[8, 9, 14]), has been described in great detail in previous papers [12]). Basically, it consists of two vertical capillaries, each closed at one end by a platinum electrode, and positioned one above the other with the open ends separated by a distance of about 14 mm. The upper and the lower tubes, initially filled with solutions at concentrations 0.75*c* and 1.25*c*, respectively, are surrounded by a solution of concentration *c*. This bulk solution is contained in a glass tank (200 × 140 × 60) mm immersed in a thermostat bath at 25 ºC. Perspex sheets divide the tank internally and a glass stirrer creates a slow lateral flow of the bulk solution across the open ends of the capillaries. Experimental conditions are such that the concentration at each of the open ends is equal to the bulk solution value *c*, that is, the physical length of the capillary tube coincides with the diffusion path. This means that the required boundary conditions described in the literature [12] to solve Fick´s second law of diffusion are applicable. Therefore, the so-called Δl effect [12] is reduced to negligible proportions. In our manually operated apparatus, diffusion is followed by measuring the ratio $w=R_t/R_b$ of resistances R_t and R_b of the upper and the lower tubes by an alternating current transformer bridge. In our automatic apparatus, w is measured by a Solartron digital voltmeter (DVM) 7061 with 6 1/2 digits. A power source (Bradley Electronic Model 232) supplies a 30 V sinusoidal signal at 4 kHz (stable to within 0.1 mV) to a potential divider that applies a 250 mV signal to the platinum electrodes in the top and the bottom capillaries. By measuring the voltages V' and V'' from the top and the bottom electrodes to a central electrode at ground potential, in a fraction of a second, the DVM calculates w.

In order to measure the differential diffusion coefficient *D* at a given concentration *c*, the bulk solution of concentration *c* is prepared by mixing 1 L of "top" solution with 1 L of "bottom" solution, accurately measured. The glass tank and the two capillaries are filled with *c* solution, immersed in the thermostat, and allowed to come to thermal equilibrium. The resistance ratio $w = w_\infty$ is measured under these conditions (with solution in both capillaries at concentration *c* what accurately gives the quantity $\tau_\infty = 10^4 / (1 + w_\infty)$.

Afterwards, the capillaries are filled each one with the "top" and the "bottom" solutions, which are then allowed to diffuse into the "bulk" solution.

Resistance ratio readings are taken at various recorded times, beginning 1000 min after the start of the experiment, to determine the quantity $\tau = 10^4/(1+w)$ as τ approaches τ_∞. The diffusion coefficient is evaluated using a linear least-squares procedure to fit the data and, finally, an iterative process is applied using 20 terms of the expansion series of Fick's second law for the present boundary conditions. The theory developed for the cell has been described previously [12].

The Taylor dispersion method is based on the dispersion of small amounts of solution injected into a laminar carrier streams of solvent or solution of different composition, flowing through a long capillary tube [13] (Figure 2). The length of the Teflon dispersion tube used in the present study was measured directly by stretching the tube in a large hall and using two high quality theodolytes and appropriate mirrors to accurately focus on the tube ends. This technique gave a tube length of 3.2799 (±0.0001) $\times 10^4$ mm, in agreement with less-precise control measurements using a good-quality measuring tape. The radius of the tube, 0.5570 (± 0.00003) mm, was calculated from the tube volume obtained by accurately weighing (resolution 0.1 mg) the tube when empty and when filled with distilled water of known density.

At the start of each run, a 6-port Teflon injection valve (Rheodyne, model 5020) is used to introduce 0.063 mL of solution into the laminar carrier stream of slightly different composition. A flow rate of 0.17 mL min^{-1} is maintained by a metering pump (Gilson model Minipuls 3) to give retention times of about 1.1×10^4 s. The dispersion tube and the injection valve are kept at 298.15 K (± 0.01 K) in an air thermostat.

Dispersion of the injected samples is monitored using a differential refractometer (Waters model 2410) at the outlet of the dispersion tube. Detector voltages, $V(t)$, are measured at accurately 5 s intervals with a digital voltmeter (Agilent 34401 A) with an IEEE interface. Binary diffusion coefficients are evaluated by fitting the dispersion equation

$$V(t) = V_0 + V_1 t + V_{\max} (t_R/t)^{1/2} \exp[-12D(t - t_R)^2/r^2 t] \tag{1.3}$$

to the detector voltages. The additional fitting parameters are the mean sample retention time t_R, peak height $V_{\max}$, baseline voltage V_0, and baseline slope V_1.

Diffusion in a ternary solution is described by the diffusion equations (eqs. 1.4 and eq. 1.5),

$$-(J_1) = (D_{11})_{\mathrm{v}} \frac{\partial c_1}{\partial x} + (D_{12})_{\mathrm{v}} \frac{\partial c_2}{\partial x} \tag{1.4}$$

$$-(J_2) = (D_{21})_{\mathrm{v}} \frac{\partial c_1}{\partial x} + (D_{22})_{\mathrm{v}} \frac{\partial c_2}{\partial x} \tag{1.5}$$

where J_1, J_2, $\partial c_1/\partial x$ and $\partial c_2/\partial x$ are the molar fluxes and the gradients in the concentrations of solute 1 and 2, respectively. Main diffusion coefficients give the flux of each solute produced by its own concentration gradient. Cross diffusion coefficients D_{12} and D_{21} give the coupled flux of each solute driven by a concentration gradient in the other solute. A positive D_{ik} cross-coefficient ($i \neq k$) indicates co-current coupled transport of solute i from regions of higher concentration of solute k to regions of lower concentration of solute k. However, a negative D_{ik} coefficient indicates counter-current coupled transport of solute i from regions of lower to higher concentration of solute k.

Extensions of the Taylor technique have been used to measure ternary mutual diffusion coefficients (D_{ik}) for multicomponent solutions. These D_{ik} coefficients, defined by eqs (1.4) and (1.5), were evaluated by fitting the ternary dispersion equation (eq. 1.6) to two or more replicate pairs of peaks for each carrier-stream.

$$V\ t = V_0 + V_1 t + V_{\max}\ t_R / t^{\ 1/2} \left[W_1 \exp\left(-\frac{12 D_1 (t - t_R)^2}{r^2 t} \right) + (1 - W_1) \exp\left(-\frac{12 D_2 (t - t_R)^2}{r^2 t} \right) \right] \tag{1.6}$$

Two pairs of refractive–index profiles, D_1 and D_2, are the eigenvalues of the matrix of the ternary D_{ik} coefficients.

In these experiments, small volumes of solution, ΔV, of composition $\bar{c}_1 + \overline{\Delta c_1}$ and $\bar{c}_2 + \overline{\Delta c_2}$ are injected into carrier solutions of composition, $\bar{c}_1$ and $\bar{c}_2$, at time t = 0.

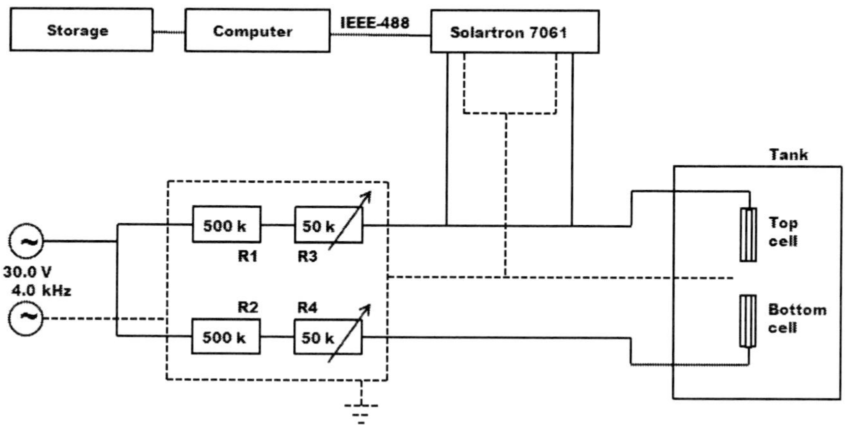

Figure 1 The open-ended capillary cell scheme. Resistances: R1, R2, R3, R4 [12].

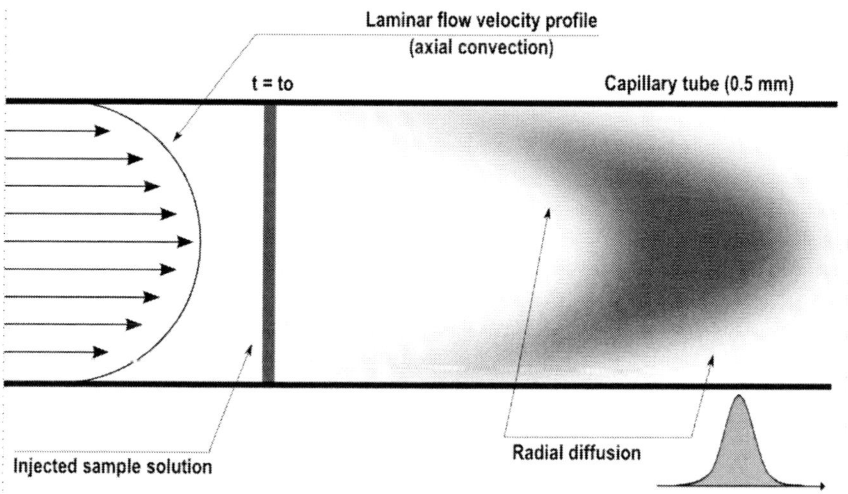

Figure 2. Schematic representation of the Taylor Dispersion technique [13].

2. Experimental Results and Discussion

Using our techniques (open-ended capillary cell and Taylor dispersion), our group has also been focused on the study of the diffusion behavior of systems involving lactose, that is the ternary system $NiCl_2$/lactose/water) [6] and pseudo-binary systems (i.e., $CdCl_2$/lactose/water [8] and $FeCl_2$/lactose

/water [9]) (Table 1). The last ones are actually ternary systems, and we really have been measuring only the main diffusion coefficients (D_{11}). However, from experimental conditions, we may consider these systems as pseudo-binary ones, and, consequently, take the measured parameters as binary diffusion coefficients, D.

Table 1 and Figure 1 present new data, that is, diffusion coefficients, D, and their respective standard deviations, S_D, of $CaCl_2$ 0.100 mol dm^{-3} in aqueous solutions of lactose (0.005 mol dm^{-3}, 0.010 mol dm^{-3}, 0.025 mol dm^{-3}, 0.050 mol dm^{-3}, 0.100 mol dm^{-3} and 0.200 mol dm^{-3}) at 25 ºC. For brevity, we will indicate as 1 and 2 (not including the solvent, component 0), the lactose and the electrolytes containing ion chloride species, respectively. D is the mean value of, at least, three independent measurements, and the error limits of our results should be close to the imprecision, therefore giving an experimental uncertainty of 1–3%.

The following polynomial in $c_1^{1/2}$ was used to fit our data by a least squares procedure (Table 1),

$$D/(10^{-9}\ \mathrm{m}^2\ \mathrm{s}^{-1}) = 1.094 + 0.328\, c_1^{1/2} - 1.464\, c_1$$
$$R^2 = 0.999 \qquad (2.1)$$

where c_1 and D represent the concentration of lactose and the diffusion coefficients of calcium chloride in different aqueous solutions containing this carbohydrate, respectively. The goodness of the fit (obtained with a confidence interval of 98 %) was assessed by the excellent correlation coefficient, R^2, and the low standard deviation (< 1 %). Moreover, the deviation between the limiting D^0 value calculated by extrapolating experimental data to $c_1 \rightarrow 0$ ($D^0 = 1.094 \times 10^{-9}\ \mathrm{m}^2\ \mathrm{s}^{-1}$) and the measured D value for solutions containing only $CaCl_2$ for c = 0.100 mol dm^{-3} ($D = 1.110 \times 10^{-9}\ \mathrm{m}^2\ \mathrm{s}^{-1}$) is also acceptable (-1.4 %).

From the difference between the diffusion measurements of $CaCl_2$ in aqueous solutions in both the presence and the absence of lactose (measured as the percentage difference, $\Delta D/D$ %) (Table 1), we can verify that the diffusion behaviour of calcium chloride in aqueous solutions at 25 ºC is affected by the presence of the lactose molecules. In fact, under thc present experimental conditions, when we have lactose at concentrations $c \leq 0.005$ mol dm^{-3}, the diffusion coefficient of $CaCl_2$ did not significantly change.

However, when the lactose concentration is increased up to 0.200 mol.dm^{-3}, the observed diffusion coefficients are lower than those obtained in aqueous solutions, reaching a maximum difference at a lactose/$CaCl_2$ ratio 2:1 (i.e., $\Delta D/D$ = -14.7 %).

The decrease of the diffusion coefficients of calcium chloride with the increasing of concentration of lactose in these media may be interpreted on the basis of the formation of new species (e.g. ion pairs) resulting, mainly, from the interactions of Ca^{2+} cations with lactose, which will lead to greater resistance to the motion through the liquid and, consequently, a lower diffusion coefficient.

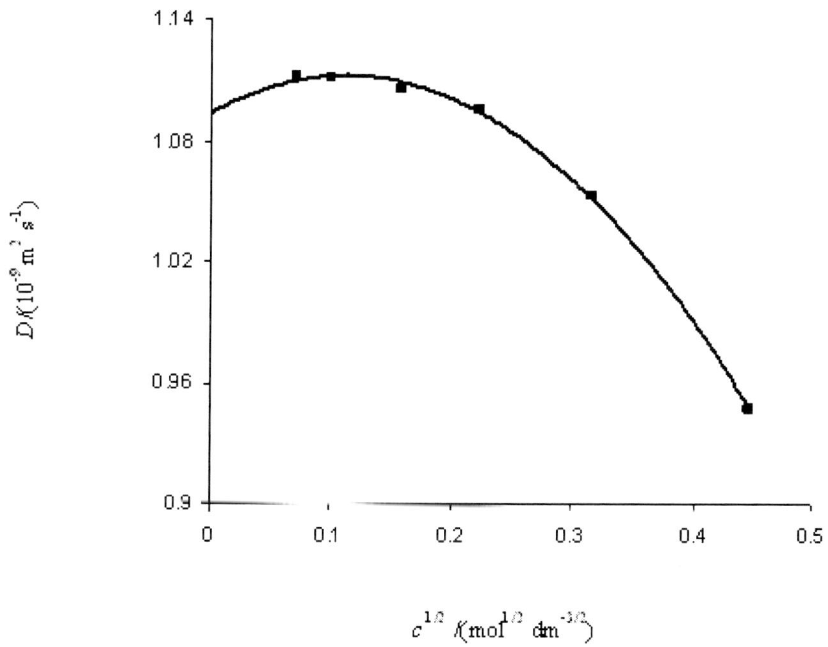

Figure 3. Diffusion coefficients, D, of aqueous lactose (c_1) + calcium chloride (c_2) solutions at 25 °C, plotted against the concentration of lactose, c_1. (■) Measured values of D (Table 1). —— Fitting curve of our diffusion coefficients of calcium chloride in aqueous solutions containing lactose at different concentrations (equation 2.1).

Table 1. Experimental diffusion coefficients, *D*, of aqueous systems containing lactose (1), and divalent metal chlorides (2) at 25 °C[a)]

Lactose (c_1) + $CaCl_2$ (c_2 = 0.100 mol dm^{-3})			Lactose (c_1) + $CdCl_2$ (c_2 = 0.100 mol dm^{-3})		Lactose (c_1) + $FeCl_3$ (c_2 = 0.05 mol dm^{-3})	
c_1 // (mol dm^{-3}) [b)]	$D \pm S_D$ // (10^{-9} m^2.s^{-1})[c)]	($\Delta D/D$) // %[d)]	$D \pm S_D$ // (10^{-9} m^2.s^{-1})[c)]	($\Delta D/D$) // %[f)]	$D \pm S_D$ // (10^{-9} m^2.s^{-1})[c)]	($\Delta D/D$) // %[h)]
0.200	0.947 ± 0.018	-14.7	0.274 ± 0.022	-69.6	1.149 ± 0.030	-22.2
0.100	1.053 ± 0.020	-5.1	0.669 ± 0.015	-25.8	1.305 ± 0.020	-11.6
0.050	1.095 ± 0.021	-1.4	0.847 ± 0.010	-6.1	1.435 ± 0.035	-2.8
0.025	1.106 ± 0.019	-0.4	0.915 ± 0.010	+1.4	---	---
c_1 // (mol dm^{-3}) [b)]	$D \pm S_D$ // (10^{-9} m^2.s^{-1})[c)]	($\Delta D/D$) // %[d)]	$D \pm S_D$ // (10^{-9} m^2.s^{-1})[c)]	($\Delta D/D$) // %[f)]	$D \pm S_D$ // (10^{-9} m^2.s^{-1})[c)]	($\Delta D/D$) // %[h)]
0.010	1.111 ± 0.028	0.0	0.931 ± 0.015	+3.2	---	---
0.005	1.112 ± 0.023	0.2	0.924 ± 0.010	+2.4	---	---
0.000	1.110 ± 0.008[e]	0.4	0.902 ± 0.028[g)]	0.0	1.476 ± 0.015[i)]	—

[a] uncertainty, u(T) = 0.01 K. [b]uncertainty, u(c) = 0.001 mol dm^{-3}. [c]D is the mean diffusion coefficient of at least 3 experiments and S_D is the standard deviation of that mean. [d]($\Delta D/D$) / % represents the deviation between the diffusion coefficients of the system (Lactose + $CaCl_2$ + H_2O), D, and the diffusion coefficients, D_{Lit}, of the system ($CaCl_2$+ H_2O) [14]. [e]Experimental value of the diffusion coefficient of aqueous solutions of calcium chloride at c_2 = 0.100 mol dm^{-3} for c_1 = 0 [14]. [f]$\Delta D/D$) / % represents the deviation between the diffusion coefficients of the system (Lactose + $CdCl_2$ + H_2O) and the diffusion coefficients, D, of the system ($CdCl_2$+ H_2O) [8]. [g]Experimental value of the diffusion coefficient of aqueous solution of cadmium chloride at c_2 = 0.100 mol dm^{-3} for c_1 = 0 [8]. [h]$\Delta D/D$) / % represents the deviation between the diffusion coefficients of the system (Lactose + $Fe_2(SO_4)_3$+ H_2O) and the diffusion coefficients, D_{exp}, of the system ($Fe_2(SO_4)_3$ + H_2O) [9]. [i]Experimental value of the diffusion coefficient of aqueous solutions of $D_{Fe2(SO4)3}$ 0.05 M for $X_1 = 0$ [9].

Also for other systems (Table 2), we observed similar tendencies concerning the decreasing of the diffusion coefficients of these divalent metal chlorides in mixtures containing lactose with an increase on its concentration, reaching, however, in these cases, more significant deviations. This behaviour can be explained if we consider the types of interactions that can occur: a) interactions between lactose molecules and the chloride anions (coulombic, etc); b) interactions between lactose molecules and the metal cations; c) changes in the interactions between the metal cation and the chloride. Also, it is probable that all these types of interaction occur, depending on the circumstances. Thus, for all systems shown in Table 1, we can say that an appreciable fraction of the transport of metal chloride may occur as a result of larger aggregates (which aggregation number may depends on the nature of the electrolyte), and consequently, that diffusion of those salts in aqueous solutions of lactose at 25 ºC is significantly affected.

Conclusion

Mutual differential isotherm diffusion coefficients, *D*, of divalent metal chlorides in aqueous solutions containing lactose have been measured, having in mind a contribution to a better understanding of the structure of those solutions, behavior of these salts in the presence of lactose and supplying the scientific and technological communities with data about these important parameters in the solution transport processes. In fact, the scarcity of diffusion coefficients in the scientific literature, due to the difficulty on their accurate experimental measurement and impracticability of their determination by theoretical procedures, allied to the respective industrial and research needs, well justifies efforts in such accurate measurements. For example, the diffusion coefficients of these electrolytes when obtained in the presence of lactose may provide transport data necessary to model the diffusion for various chemical and pharmaceutical applications.

Acknowledgments

Financial support from FCT (FEDER)-PTDC/AAC-CLI/098308/2008 and PTDC/AAC-CLI/118092/2010 is gratefully acknowledged.

References

[1] M.L. Swartz, R.A. Bernhard and T.A. Nickerson (Interactions of Metal Ions with Lactose), *J. Food Science* 43, 93-97 (1978).

[2] Y. Sano and S. Yamamoto (Mutual diffusion coefficient of aqueous sugar solutions), .J. *Chem. Eng. Jpn.*, 26, 633-636 (1993).

[3] P.K. Banipal; T.S Banipal; B.S Lark and J.C. Ahluwalia (Partial molar heat capacities and volumes of some mono-, di- and trisaccharides in water at 298.15, 308.15 and 318.15 K.), *J. Chem. Soc.*,Faraday Trans.93, 81-87(1997).

[4] G.R. Moran, K.R Jeffrey,J.M Thomas and J.R Stevens (A dielectric analysis of liquid and glassy solid glucose/water solutions), *Carbohydr.Res.*32, 573-584 (2000).

[5] A.C.F. Ribeiro, O. Ortona, S.M.N. Simões, C.I.A.V. Santos, P.M.R.A. Prazeres, A.J.M. Valente, V.M.M. Lobo and H.D. Burrows (Binary Mutual Diffusion Coefficients of Aqueous Solutions of Sucrose, Lactose, Glucose and Fructose in the Temperature Range 298.15 to 328.15). *J. Chem. Eng. Data* 51, 1836-1840 (2006).

[6] A.C.F. Ribeiro, J.C.S. Gomes, M.C.F. Barros, V.M.M. Lobo and M.A. Esteso (Diffusion Coefficients of Nickel Chloride in Aqueous Solutions of Lactose at 298.15 K and 310.15 K). *J. Chem. Thermodynamics* 43, 270–274 (2011).

[7] A.C.F. Ribeiro, M.B.B.J.R, J.C.S. Gomes, V.M.M. Lobo and M.A. Esteso (Diffusion of Calcium Gluconate in Aqueous Solutions of Lactose at 298.15 K), *Food Chemistry* 126, 1186–1189 (2011).

[8] L.M.P. Veríssimo, J.C.S. Gomes, C.Romero, M.A.Esteso, A.J.F.N. Sobral and A. C.F. Ribeiro (Interaction between Lactose and Cadmium Chloride in Aqueous Solutions as Seen by Diffusion and ab initio Calculations). *J. Chem. Thermodynamics* 61, 79-82 (2013).

[9] L.M.P. Veríssimo, T.I.C. Valada, A.C.F. Ribeiro, A.J.F.N. Sobral, V.M.M. Lobo and M.A. Esteso (*Effect of Lactose on the Diffusion of Ferric Sulphate in Aqueous Solutions at 25 ºC*), *J. Chem. Thermodynamics* 59, 135–138 (2013).

[10] R.A Robinson and R.H. Stokes. *Electrolyte Solutions*. 2nd Ed., Butterworths, London, 1959.

[11] H.J.V Tyrrell, K.R. Harris. *Diffusion in Liquids*. 2nd Ed., Butterworths, London, 1984.

[12] J.N. Agar, and V.M.M. Lobo (*Measurement of diffusion coefficients of electrolytes by a modified Open-Ended Capillary Method*), J. Chem. Soc., Faraday Trans. I 71, 1659-1666 (1975).

[13] R. Callendar and D.G. Leaist (Diffusion coefficients for binary, ternary, and polydisperse solutions from peak-width analysis of Taylor dispersion profiles), *J. Sol. Chem*. 35, 353-379 (2006).

[14] A.C.F. Ribeiro, M.C.F. Barros, A.S.N. Teles, A.J.M. Valente, V.M.M. Lobo, A.J.F.N. Sobral and M.A. Esteso (*Diffusion Coefficients of Calcium Chloride in Aqueous Solutions at 298.15 K and physiological temperature (310.15 K). Electrochim. Acta*.54, 192-196 (2008).

Index

C

D

E

F

G

H

O

P

Q

R

S

T

U